Am I Making Myself Clear?

Am I Making Myself Clear?

A Scientist's Guide to Talking to the Public

Cornelia Dean

HARVARD UNIVERSITY PRESS
Cambridge, Massachusetts
London, England
2009

Printed in the United States of America

Library of Congress Cataloging-in-Publication Data

Dean, Cornelia.
Am I making myself clear? : a scientist's guide to talking to the public / Cornelia Dean.
p. cm.
Includes bibliographical references and index.
ISBN 978-0-674-03635-2 (alk. paper)
1. Communication in science. 2. Science news.
3. Scientists—Vocational guidance. I. Title.
Q223.D43 2009 2009019750
501'.4—dc22

This book is dedicated to the scores of researchers
who help me and other journalists
enlighten our fellow citizens
about the marvels, perils, and
promise of science and technology.

"Give light and the people will find their own way."

—E. W. Scripps Company,
founder of the Scripps Howard newspaper chain

Contents

Am I Making Myself Clear?

I

An Invitation to Researchers

I work as a science journalist. That is, I pay attention to interesting and important developments in science and engineering, talk to the researchers who uncover them, learn about the ideas behind them, and then communicate this information to the public as engagingly as I can.

I like this work not just because research is fascinating, though of course it is, but also because scientists and engineers are interesting. Typically, they are passionate about their work—and passion is an attractive quality.

I did not begin my career as a science writer. As

a young journalist, I covered school board meetings, city council deliberations, crimes, plane crashes, auto wrecks, and politics. I even reported on a scandal or two. For a brief period, I worked in Washington covering Congress and later worked there again, as an editor whose portfolio was domestic policy.

These subjects don't fascinate me the way science does, but covering them is important. When people learn what their school board is doing about teacher salaries, or what the zoning board thinks about a planned apartment complex, or even how their senator votes on taxes or trade policy, they gain information they need as citizens. The same is true of science and technology. Nowadays, people cannot fully function as citizens unless they understand developments in these fields.

Jon Miller, a researcher at Michigan State University who studies public attitudes toward and knowledge of science, compares these times to colonial New England, when (white, male, property-owning) citizens conducted town business by holding votes in town meetings. When Miller studied records of the issues that came up at those meetings—building fences, cutting roads, digging wells, or banning dancing, say—he concluded that people could form opinions about the issues and vote intelligently on them even if they were illiterate, as many voters were.[1] That is no longer the case.

As Richard Gallagher, editor of the *Scientist,* put it in an editorial, "We need an informed public if social policies are to be decided on reasonable and rational grounds. Everything from the future of health care and how it's paid for, to taxation on fuel, could benefit from a wider appreciation of the wider science. Not to mention evolution versus intelligent design and stem cells."[2]

Add to those issues climate change, improving our aging infrastructure, the protection of endangered species, weapons of mass destruction, health care policy, the ideal objectives of the space program, and a host of others. People cannot form intelligent opinions on these subjects—they cannot be intelligent *voters*—unless they understand the technological questions that underlie them.

But people do not understand science. In survey after survey, they display ignorance, superstition, and patterns of irrational thinking so persistent they seem to be hardwired into the human brain. Unfortunately, in politics, business, and elsewhere, there are plenty of people prepared to profit from these weaknesses.

They know that Americans respect science. According to a survey by Research!America, an organization that supports science research, 87 percent of Americans say scientists are people of "very great" or "considerable" prestige, the highest of any job category surveyed. (The comparable figure for journalists is 46 percent.)[3]

So people with a political point to make attempt to cloak their arguments in the rhetoric of science, even if they are twisting the facts, and even if the issue they are arguing about is not a question that science or engineering can answer.

Climate is a good example. The central climate-change question—is human activity altering the chemistry of the atmosphere in ways that threaten dire consequences?—was long ago answered in the affirmative. The only live issues are values issues, like whether people living today have the right to continue to enjoy a carbon-fueled good life at the expense of generations to come, or policy issues, like whether a carbon tax or a cap-and-trade program is a better approach to the problem.

But instead of confronting these issues head on, opponents of action delayed it for decades, arguing that scientific findings on climate change were too flimsy to warrant action that might disrupt the economy. Only if policymakers—and voters—understand enough science to see through these arguments can they confront the underlying issues productively.

Anyone who pays attention to the news can cite numerous other examples. There are those who argue that the administration of George W. Bush turned this kind of deception into a high art. And it's true, his administration was exposed again and again as having manipulated science to suit political aims by misstating facts, magnifying scientific uncertainty, suppressing inconvenient truths, targeting scientists who challenged administration misstatements, and stacking advisory committees and technical agencies with people notable more for their political or religious views than their technical expertise. The Bush administration even undermined the scientific enterprise itself, as when Bush declared that a form of creationism should be taught along with the theory of evolution in the nation's biology classes.

But right-wing politicians like Bush and his associates are hardly the sole practitioners of this art. Though leading Democrats today seem more likely than leading Republicans to have some understanding of science—President Carter trained as a nuclear engineer, for example, and as vice president, Al Gore understood the Internet, even if he did not actually invent it—the left wing is not innocent of the charge that it twists science to suit facts. For example, while Gore's 2006 movie *An Inconvenient Truth* brought the issue of climate change vividly to the attention of millions of Americans, it walks right up to the line of scientific consensus on climate and, some would argue, occasionally puts a toe over the edge.

Environmental and health advocacy groups have made their own unbalanced claims. Drug companies, doctors, and patient groups can commonly be found asserting that particular medical treatments have benefits that, at a minimum, have yet to be proved. Or they assert as proved, as some have with mercury-based preservatives in vaccines and autism, links that do not exist.

At the same time, there has been an erosion of the federal government's technical infrastructure in the very organizations that people—and public officials—might once have turned to for impartial expert advice. The congressional Office of Technology Assessment has been abolished. In the Bush administration, the office of White House science adviser was downgraded in influence and prestige—something President Obama reversed. Since the end of the cold war, scientific expertise has drained away from a host of government agencies once deservedly proud of their technical chops.

So we find ourselves in a society in which science is often misrepresented and arguments about values are often presented as if they are legitimate scientific disputes. As a result, people can become so disenchanted and confused that they turn away from science and engineering expertise as a source of guidance they can rely on in the voting booth.

As a science journalist, of course I believe better journalism would help turn this situation around. But the news business is in a period of turmoil. The Internet is drawing more and more advertising revenue from newspapers, whose circulation is falling as readers who once would have been subscribers spend more and more time getting their news online. And though news organizations are spending more and more resources online, they have so far failed to make money there the way they did with paper.

Meanwhile, the proliferation of cable television channels and Internet sites has fragmented and shrunk the audiences of once-mighty television networks. Changes in Federal Communications Commission regulations have led to greater concentration of media ownership and the virtual elimination of so-called public interest requirements, further encouraging reduced spending on news.

As newsroom managers look for ways to cut costs, science segments and even science pages are biting the dust. Though there is still some emphasis on engineering or biotech news when it seems to have short-term implications for the marketplace, science reporters are being shifted to other beats or fired outright, leaving fewer and fewer journalists with less and less support to cover topics whose importance and complexity only grows. And it would be highly unrealistic to expect things to change for the better any time soon.

I am happy to say, however, this is not universally the case. At the *New York Times,* where I work, science journalism has long been highly valued. When I was the newspaper's science editor, its management routinely referred to our science and health coverage as one of the jewels in the newspaper's crown and (from a less high-flown perspective) one of the features that differentiated us from our competition and drew and retained readers. I think our management was right, and I often wondered why other news outlets did not adopt our approach.

So long as they don't, though, journalists will find themselves struggling to cover complex technological issues with fewer and fewer resources. Who can help us? Researchers.

But researchers are taught not to spend time on anything but research. For most, the demands of their work mean they already have enough on their collective plate. As one eminent climate expert once said to me, in the midst of complaining about the poor quality of science journalism, "Every minute away from the bench is time wasted." Also, many researchers scorn the mass media as an arena where important work is all too often misrepresented or hyped. And they are unlikely to achieve tenure, win a grant, or be offered a promotion because they have been mentioned in the news. As Donald Kennedy, president emeritus of Stanford, told an audience at the American Academy of Arts and Sciences, "In iconic PhD-granting departments, students are seldom asked to work on their communication skills." In the rare instances when they *are* offered training in communicating about science, he added, "it is a commonly heard instruction that they should focus on the thesis" instead.[4]

In fact, Kennedy said, if students express an interest in communicating research findings to the public they will ordinarily be treated to a lecture on what Kennedy called "the dangers of being Sagan-ized." He was referring to the so-called Carl Sagan effect, named after the Cornell astrophysicist who was denied a place in the National Academy because, it is widely believed, his public television series, *Cosmos,* was all too successful with the general public.

This pattern of thinking must change. But until it does, it is no surprise that many researchers have a simple response to inquiries from reporters: they ignore the calls. This practice is unfortunate, because they could do much to help reporters. And if they participated more in the public life of our nation, if they dropped their institutional reticence and let their voices be heard beyond realms of scholarly publication, they could do more than that. They could inject a lot of rationality into our public debates.

So my goal in this book is to help that effort by discussing some of the barriers to good public understanding of science and technology, describing the journalistic landscape in which public discussion of these issues takes place, and identifying ways researchers can participate in this discourse and raising the odds that the experience will be useful.

This book is not a work of scholarship. Still less is it imbued with technical expertise. It cannot be a thorough guide, even for this landscape. But I hope it will lead readers to know what kinds of things they can expect if they reach out to the public, and will offer useful hints for improving their interactions with policymakers, the public, and journalists like me. More important, I hope it will convince researchers that communicating their work and the work of others to the lay public is important for society—and a valuable use of their time.

Of course, no book can turn someone into a Carl Sagan or Rachel Carson or Stephen Jay Gould. But with luck, researchers who read this book will learn how to speak to the public, deal with the media, describe their own work to a lay audience on paper, online, and over the air. They will learn a bit about making a difference in legislation, litigation, and other areas. I will discuss the public's knowledge of and attitudes toward science and technology, the landscapes of journalism generally and journalism on these issues in particular, how researchers can be good sources for journalists, how to participate in the larger life of the nation—in litigation, policymaking, politics, and elsewhere.

If you are a researcher, I hope you will find the book's lessons useful and apply them in a career filled with public engagement. If you decide against this kind of effort, I hope you will at least support those who *do* engage and greet them with praise and not derision.

This book is far from the only effort that has been made to address this problem. The Union of Concerned Scientists covers some of this ground in its publication, *A Scientist's Guide to Talking with the Media.* The American Association for the Advancement of Science has produced *Working with Congress,* a guide for scientists and engineers. It also runs internship programs that put scientists and engineers into the newsrooms of prominent news outlets and into the offices of federal officials and members of Congress. Organizations like the Packard Foundation and the Pew Foundation have supported programs to train a few elite scientists every year in taking a larger role in the nation's public discourse. A few universities are starting to address these issues in training graduate students in science. (In fact, this book emerges from the seminars I teach at Harvard.)

Of course, there is no particular reason to accept my suggestions as gospel. I hope, though, that this book will get you thinking about the proper role of researchers in the wider world, and how you can do your part to fill it usefully. If you do that, you will help make things better for all of us.

2

Know Your Audience

A few years ago, I was one of about a hundred participants, most of us science journalists or would-be science journalists, in an experiment on extrasensory perception. The experimenter was Josh Tenenbaum, a cognitive scientist at the Massachusetts Institute of Technology. Tenenbaum had a penny—a fair coin, he told us. He would flip it five times. Each time he would send "mind rays" out into the room telling us whether the coin had come up heads or tails. All of us would try to receive his signals and then write down what message we received. At the end of the experiment, we would know who had extrasensory perception and who did not.

He flipped. We wrote. He flipped again, we wrote again. After the final flip, he asked how many of us sensed that the penny had turned up heads-heads-tails-heads-tails. About a third of us, including me, said that we had. He asked how many thought the penny had landed tails-tails-heads-tails-heads (the obverse). Another quarter of the group raised their hands.

Then he asked how many of us thought the penny had landed on heads every time, or tails every time. None of us had come up with HHHHH or TTTTT, even though, statistically, they were just as likely as HHTHT or TTHTH.

Then he told us, of course, there were no such things as mind rays and that the "experiment" had nothing to do with extrasensory perception. Instead, it was yet another demonstration of how people have wrong ideas about randomness—about statistics. We know what randomness looks like—it looks like HHTHT or TTHTH—but we are wrong.[1]

Tenenbaum had pointed out one of the three major deficits ordinary Americans exhibit when they try to think about science: we don't reason probabilistically, we don't understand the scientific method, and we don't know much.

By now it is a truism that Americans don't much like studying science. In fact, when the National Research

Council, the research arm of the National Academy of Sciences, surveyed biology students at the beginning and the end of the tenth grade, they found they were less interested in science at the end of the year than they had been at the beginning. That is, the experience of studying tenth-grade biology, usually the first serious science course students encounter, was not encouraging them to explore the wonders of nature or the possibilities of engineering. It was turning them off.

Volumes have been written about why this is so. Blame is variously ascribed to poor teaching; teaching science in the wrong order (biology-chemistry-physics rather than the reverse, which would make more sense scientifically); our long anti-intellectual tradition (even in the *New York Times* newsroom I have heard people almost brag of their innumeracy); laboratory exercises that are more like following cookbook recipes than journeys of discovery; and, my personal pet hate, in college the ubiquitous introductory science courses designed not to draw students into majoring in math or engineering or chemistry, but to drive them away.

For our purposes, it is enough to know that many people who studied science and math in high school or even college remember the experience unhappily as a time in which they were alternately bored or humiliated. So it is not surprising that many Americans don't know that the earth goes around the sun and takes a year to do it, that molecules are larger than atoms, or that all plants and animals, genetically engineered or not, have genes.

Perhaps as a result, people don't follow science as well as they might. According to the Research!America's Bridging the Sciences survey, when Americans were asked to name a living scientist, 74 percent drew a blank. Stephen Hawking was named by 8 percent of participants, but no one else drew more than 1 percent. One of the "living scientists" named by those surveyed was Albert Einstein, who died in 1955. Another was Robert Jarvik, in the news at the time because his endorsements of the anticholesterol drug Lipitor had come into question.[2]

According to the same survey, when people were asked if they could name any nearby institutions, companies, or organizations that engaged in research, many—including 40 percent in Massachusetts, 53 percent in California, and 57 percent in Texas—had nothing to say.

More important than ignorance of technical facts is ignorance of the way science works. Science looks to nature for answers to questions about nature, tests those answers through experiment and observation, and considers the results provisional—valid only until new experiments or observations challenge them.

The public's ignorance of these features of science

helps explain how otherwise savvy people can think that creationism or its ideological cousin, intelligent design, is appropriate for a science classroom. They don't understand that a theory that relies on the action of a supernatural entity is, by definitions that have prevailed since the Enlightenment, not science.

Our reasoning is equally flawed. Given examples, we generalize. Given effects, we infer causes. Instead of viewing correlation for what it is—an opportunity to hypothesize about causation—we assume it *proves* causation. And for us, vivid anecdotes mean more than piles of data.

We do not reason probabilistically—and not just when it comes to coin-tossing. We do not understand, for example, that if a phenomenon occurs widely and randomly in a large population, there will be places where it will seem to have occurred in a cluster—a cancer cluster, say. If such clusters did not exist, the distribution would be too uniform to be random.

H. G. Wells wrote decades ago that it was not enough for citizens in a technological society to know how to read and write. They had to know statistics too. It is tempting to imagine what might happen if the nation's scientists and engineers got together and insisted that the Educational Testing Service, which administers the SAT and other standardized tests, devote more space on the SATs to statistics. My guess is the benefits would be widespread.

In any event, these patterns of faulty thinking are why Robert Park, a physicist at the University of Maryland and representative in Washington for the American Physical Society, calls the human brain "a belief engine" that "knows nothing of the laws of probability."[3] He also notes that, in the United States, anyway, belief in that which reason denies is regarded not as delusion, but rather as a sign of steadfastness and courage. In contrast, "Prove it!" skepticism, the central characteristic of scientific inquiry and engineering's precision, is often portrayed as carping.

At the same time, ignorance of how science works can lead people to distrust science. Many people believe that research leads to the discovery, one after another, of a cavalcade of facts that will stand forever. They do not understand that, instead, research is an ungainly mechanism that moves in fits and starts and that its ever-expanding path of knowledge is complicated by blind alleys and fruitless detours. When researchers challenge each other, as they will on topics as weighty as climate or as trivial as coffee drinking, ordinary Americans do not admire the intensity of their search for truth; rather, they think "those scientists are all over the place—they must not know what they are talking about."

In my view, it is this ignorance of how research is done, as much as ignorance of scientific or technical facts, that leads so many Americans to embrace creationism, astrology, and UFOs.

Statistics on this point vary according to how questions are asked, but the National Science Foundation, in its 2008 survey of attitudes toward and knowledge of science, found that a minority of Americans—about 45 percent—accept the theory of evolution, a percentage that has held more or less steady for years. The survey also found that most of our fellow citizens assert that astrology is at least "somewhat" scientific and that a majority believe that some people possess psychic powers.[4] Park says this is why his local bookstore stocks three times as many books on UFOs as it does on science. It may also explain why practically every daily newspaper publishes a daily horoscope—while few devote regular coverage to science.

In the era before Albert Einstein and Max Planck, the biologist T. H. Huxley could define science as "nothing but trained and organized common sense." I suppose that definition still holds, but in a world of relativity and quantum mechanics, it doesn't feel like it. Common sense notions about reality have been shattered. The mere fact that something seems preposterous is no longer evidence of anything. And as research grows more arcane, more strange, and more specialized, only scientists or engineers are qualified to judge the work of their fellow practitioners—further separating them from the rest of us.

Questions of risk illustrate the kind of irrationality people bring to scientific discussions. Theoretically, risk is defined by a formula: (chances of exposure / occurrence) × (degree of harm). But people do not think about risk in terms of formulas. That is one reason why people in industrialized countries, who are freer from risk of real harm than just about any people who have ever lived, worry more about it now than ever before, and why questions of risk are often the central, or even the only, issues in science-related policy discussions.

Perhaps, some researchers say, that is because we are more and more dependent on new technologies with powerful effects—intended and unintended—whose risks we cannot calculate. Also, we are affluent, so we have more to lose. And we increasingly distrust institutions we used to rely on, such as religious authorities, business leaders, and government officials, especially as debates about risk are increasingly politicized.

There is abundant research on factors that differentiate the things people worry about from the things they *ought* to worry about but accept with hardly a qualm.

For example, we are more likely to be afraid of things we cannot control or cannot see, and things whose forces are artificial, unfamiliar, or new.

It is factors like these that make us, overall, more afraid of flying than driving, even though driving is far more dangerous, or more afraid of pesticide residues on our fruits and vegetables than the health risks of not eating lots of fruits and vegetables. When mad cow disease emerged in France, some commentators pointed sourly to people voicing alarm about the threat even as they puffed away on their Gauloise cigarettes. It is also probably safe to say that far more people worry about contracting mad cow disease than developing arteriosclerosis from eating beef.

So it is no surprise that when it comes to risk perception there are striking differences between the technologically adept and the lay public. In 1990, when the Environmental Protection Agency (EPA) surveyed its scientists and members of the public on environmental risks, the survey found members of the public were most worried about oil spills, hazardous waste, and releases of radioactive materials. EPA scientists put those concerns at the bottom of their list, as their effects are relatively limited and short-lived. Their hit parade: global warming, destruction of the ozone layer, destruction of habitat, and loss of biological diversity. They listed hazardous chemicals and pesticides as concerns, but only for the people who worked with them.

In a democracy, though, public opinion counts, so problems that offer relatively small risk yet worry more of the public are pursued relentlessly, while genuinely risky issues are often left alone. Presumably, issues like risk perception would be less troublesome if people were better educated about science and technology. But improving education can take us only so far. Most of us learn most of the science we need to know as citizens *after* we leave school.

No one over thirty today learned about stem cells in high school or college—the field was too new. Similarly, antimissile defense, privatizing fisheries, and even climate change were not in the curriculum when most American adults were in high school. Suppose someone asked you to identify the scientific or engineering issues children in high school today will confront as voters in twenty or thirty years. Could you do it? No matter what you may think, the answer is no. The science we consider as citizens is not the facts collected in the textbooks we read in school, but rather the cutting-edge findings that confront society with new issues.

A technologically literate public must have some understanding of the scientific method and some knowledge of statistics. Members of this public must know the difference between value judgments and statements of fact, between hypothesis and theory (in the scientific sense). They must be able to make some kind of sense of research findings when they are reported.

But even if they achieved a utopian level of understanding, they will still need someone to report the findings—and explain what they mean.

Who will do that? Journalists.

3

The Landscape of Journalism

The conventional (mainstream) journalism that people like me grew up on emerged around the turn of the twentieth century, when newspaper publishers embraced the idea of presenting the news "without fear or favor," a motto of Adolph Ochs, founder and patriarch of the family that controls the *New York Times.* The ideals of the profession—and it was a novel idea to refer to it as a profession—were independence, objectivity, fairness, and an adversarial relationship with those in power. The goal, as an old saying had it, was to "afflict the comfortable and comfort the afflicted."

But there was another saying, the motto of the

Scripps Howard newspaper chain, that I like better: "Give light and the people will find their own way." Naive as it may sound, that motto sums up my reasons for spending a career in journalism.

In much of the twentieth century, high-end practitioners of the craft of careful beat reporting and investigative journalism found themselves working to open doors and shine lights on all segments of society. They formed in effect a kind of shadow government. In areas with responsible newspapers, there would in theory be a reporter almost everywhere the government was at work—in the courthouse, the police station, the town hall, the board of education, the state legislature, the Congress, regulatory agencies, the White House.

It was journalists practicing in this professional environment who uncovered Watergate, published the Pentagon Papers, and discovered government corruption of all kinds. Along the way, they won the support of the public and, crucially, the Supreme Court.

Beginning in the 1960s, the Supreme Court expanded the First Amendment's prohibition of government interference with the news media. Far from merely prohibiting government "prior restraint"—blocking journalists from giving their news—the Court embraced the idea that free and robust news media were crucial for democracy. As a result, it protected news media from libel suits and other litigation unless the journalists' actions were dishonest or flagrantly negligent.

The result is a media environment unique in the world, one in which the press is unusually free. Libel laws are construed to give responsible journalists a pass if, through an honest mistake, they defame a government official or other person involved, willingly or not, in an issue of public importance. The Court reasoned that penalizing journalists for honest mistakes would stifle their reporting.

And so, in the United States, there is no national union of journalists to which we must belong, no licensing authority, no nothing. In effect, people are journalists if they declare themselves to be journalists.[1] As a journalist, I embrace this press freedom. But it comes at a price. Innocent people who suffer from journalists' innocent mistakes may have little recourse. The absence of licensing requirements for reporters means that people can report on whatever they want, whether they know anything about it or not. The result can be incompetent reporting, especially in coverage of highly complex subjects—like science and engineering.

Also, the model of full, fair, accurate reporting does not do well when it lapses into the he-said-she-said game, a particular problem with controversies in technical arenas. So mainstream journalists find themselves wondering if the kind of journalism they do is adequate to maintain the functioning of a democratic society. Some are trying new paradigms.

The kind of commentary Fox News offers is one new approach. Is it a success? I would say no. At least one survey has found that the more people watch Fox News, the more they are misinformed. Another new paradigm is so-called citizen journalism, in which news outlets adopt a didactic posture to educate the public about the issues they face, rather than simply reporting the news. Is this format a success? Again, I would say no. The results are too often dull.

Some publications—the *St. Petersburg Times* in Florida, the *New London Day* in Connecticut, and the *Anniston Star* in Alabama, to name three—are owned by foundations. They must make enough money to survive, but they don't have to satisfy the demands of Wall Street. Is this the journalism model of the future? Possibly.

And recently some journalists at the *Star Tribune* of Minneapolis, once one of the nation's leading dailies, organized an online, nonprofit daily newspaper. They made the decision after round after round of cutbacks and buyouts seriously diminished the newspaper's reporting capacity.[2] Newspapers in Detroit, Seattle, and elsewhere are switching to publishing all or part of their work online.

All are varieties of conventional journalism. But will conventional journalism remain the paradigmatic format? If not, what will replace it?

One answer might be personalized Web sites or Weblogs—blogs. According to Technorati, a Web site that indexes blogs, by 2008, 80,000 new blogs were being created every day. The site says the size of the blogosphere doubles every five and a half months.

At their best, blogs can create communities of like-minded people and be a check on the mass media. It is not unusual anymore for people in the mass media to find themselves chasing a story that broke first on a blog. And in an ideal world, blogs might be a way of expanding "the ephemeral daily conversation." Newspapers like the *New York Times* are betting on it, with blogs they are establishing on their own Web sites.

But given their vast number, how many blogs can find a wide readership? The answer appears to be very few. And maybe that's a good thing. Unless they have high editing standards, some blogs may end up cluttered with rumors, conspiracy theories, weird ideas, and so on. And ordinary readers may be hard-pressed to differentiate between content that is reliable and content that is not.

This confusing situation is why I hope the proliferation of information sources will ultimately bring people back to mainstream media outlets, on paper or online, in search of someone to help them wade through the morass of content and collect or "aggregate" the material they need to know about. Anyway, the biggest problem for journalism now is not that there are too many sources of news, but rather that too few people are paying attention to the news.

This problem is not new. Measured by what people in the news business call "penetration," the percentage of households subscribing to a daily paper, newspaper readership in the United States peaked in the late 1920s. It began falling with the advent of radio (sales of sheet music and other impedimenta of home music-making did too) and continued to drop with the arrival of movies and then television. Today, though, the decline is accelerating, and with it the decline in revenues. True, readership is increasing online, but not enough. And revenue is not yet following it there—not enough, anyway.

According to a recent report from the Shorenstein Center for the Press, Politics and Public Policy at Harvard, young people (teenagers and young adults) are far less interested in news than older people, and the gap between them is growing.[3]

The survey did not rely merely on participants' assertions about their news-reading and news-watching habits, because people habitually exaggerate how much attention they pay to the news. Instead, it tested their replies by asking them questions about important stories of the day. The results were depressing. The researchers found that when people say they regularly pay attention to the news, they really mean "sometimes."[4]

Meanwhile, journalism as a business is in deep trouble. As media conglomerates acquire major news organizations, there are fewer individual outlets with individual community voices. And as Wall Street looks with increasing unease at the prospects of the mass media, investors start to treat media stocks like junk bonds, demanding ever higher returns. It is not unusual today for a newspaper to be required to produce returns on equity of 20 or 25 or 30 percent—unheard of generally in the industry, where a return of 7 percent is considered average. (In 2006, when it had what were until then the greatest profits of its corporate life, even Exxon Mobil's return was less than 20 percent.)

How can a news organization achieve such returns? By cutting its budget for news gathering, making it harder for journalists to find the time, money, and space they need to put their stories into the newspaper or on the air. Often, science coverage is an early victim.

There is another problem for coverage of scientific research, one less easy to solve. It is the poor match between what researchers do and what ordinary journalists think of as news.

I say this even though, over the years, practitioners and scholars of journalism have rarely agreed on how to define "news." Joseph Pulitzer, the nineteenth-century founder of a chain of newspapers who later endowed the Pulitzer Prize, print journalism's highest honor, said news comprised entertainment, public service, and information, a definition vague enough to cover practically anything. In his autobiography *City Editor,* published in 1934, Stanley Walker of the *New York Herald* offered an alliterative definition for news: "women, wampum and wrongdoing."[5]

When he spoke to students in a seminar I was teaching at Harvard, Bill Blakemore, a longtime correspondent for ABC television news, defined news as anything that, once people learned about it, made them think "I am glad I learned that." I like this definition the best.

But in the end, all definitions are at once too limited and too inclusive. We are left to paraphrase Supreme Court Justice Potter Stewart who, on the subject of pornography, said famously that while he could not define it, he knew it when he saw it. This is what journalists call "news judgment"—an instinct for what people are going to want to know about, what they will be talking about, and what they will expect to see in your publication or on your broadcast or Web site.

There are, however, a few attributes that tend to make one item or event more "newsworthy" than another:

Extent. Something that covers a wide area or affects a large number of people—a heat wave, say—is more newsworthy than the same phenomenon, narrowly felt.

Intensity. Something whose effects are deeply felt—a killer heat wave—is more newsworthy than something that is barely noticed, even if it is barely noticed by many people and deeply felt by only a few.

Consequence. If a thing or event is going to have major repercussions—if it is going to change medical practice, move the stock market, or close a highway, it is more likely to make news.

Eminence or celebrity—the importance or fame of the people or institutions involved. As the old saying has it, "Names make news."

Proximity, or the good old "local angle." This is one

reason why press releases for journals like *Science* and *Nature* identify the work sites of researchers whose papers they publish—they know news outlets love reports about people in their area.

Timeliness. The French writer André Gide once described journalism as anything that will be less interesting tomorrow than it is today. That is true almost always. It's one reason reporters are generally more interested in following stories they get first (scoops) than in chasing scoops produced by their competition.

Novelty. Obviously.

Human interest. Ditto.

Currency. Currency is what you have when some subject that has been around forever suddenly becomes the thing everyone is talking about. Poverty enjoyed a boom like that in the 1960s; climate change is enjoying one at the moment.

As you can see, the relative importance of a given piece of news varies from place to place, from time to time. Also, there is only one page one (no working journalist actually calls it the "front page"), and there is only half an hour in the average evening news broadcast (about eighteen minutes, after you deduct commercials and promos). Even as these restrictions give way in the age of the Web, most news organizations are still limited by their own reporting and editing budgets and by the attention span of their audience. In short, space—and readers' and viewers' time and attention—is at least somewhat limited.

And when you report the news, you must answer what journalists call the "so what?" question. Philip M. Boffey, one of my predecessors as science editor of the *Times,* used to remind his staff that what readers want to know is, why are you telling me this, and why now?

There are news outlets, like the *New York Times,* where the answer "because it's cool" is enough to get something into the paper, even onto page one. Journalists working elsewhere have to present a stronger rationale for giving the story time, space, and money.

Television and radio offer further complications. Though some people download or record all or part of their regular programs, most still cannot be relied upon to replay a program if something puzzles them. A TV or radio news account must be immediately clear on first hearing. Thus complex stories are hard to deal with on the air. They must be very well told. Stories about developments in science or engineering often fall into this class, and many news directors have to be talked into tackling them.

Also, a focus on newsworthiness can cause journalists to pay inadequate attention to events and trends of genuine long-term importance. A few years ago, the *Economist* asked its readers to identify the most important news events since the magazine's founding in 1850. Here is the list they came up with:

1. The vast change in the status of women
2. Freud and the development of psychoanalysis
3. Darwin's theory of evolution
4. The development of communism
5. Fascism and the rise of totalitarian dictatorships
6. Invention of the automobile
7. Electricity and its offshoots (light, telegraph to television, movies)
8. The end of slavery on the basis of color
9. The end of monarchy as a form of government
10. The conquest of space

You might not agree with their choices. If I were compiling such a list, there are things I would lose and things I would add. What's interesting about the list, though, is not its individual components but a quality most of them share: they are not the discrete, one-time-only, "the police said today" kinds of events most people think of when they think "news." Many of them are slow-moving, incremental processes that came to some sort of fruition over a period of years, hardly discernable while the process was going on.

As my *Times* colleague Andrew C. Revkin puts it, you are unlikely ever to pick up a newspaper or log on to a news Web site and see the headline GLOBAL WARMING BREAKS OUT. It is not that kind of a story.

There is not a lot you as a researcher can do to alter these systemic problems of journalism. But you can be conscious of them and compensate for them.

4

Covering Science

In 1978 New York City was in deep financial trouble, and the *New York Times* along with it. In typical fashion, the newspaper decided to expand.

At that time, the daily paper had two sections, one headed by page one and the other a "metro front," featuring news of New York City and its region. The new paper would have four sections, fronted, respectively, by page one, metro, business coverage, and a topic that would vary every weekday. The candidate subjects were sports for Monday, food and décor for Wednesday and Thursday—traditional days for coverage of those subjects—and "weekend," a compilation of movie reviews, concert listings, and the like, for Friday. The question was: what about Tuesday?

The executives who ran the newspaper's business side lobbied for fashion, as a potentially potent generator of advertising revenue. But Abe Rosenthal, then the paper's executive editor, insisted that it be a section with intellectual bite. Thus was born Science Times.

For a while, the section was not particularly popular with advertisers. The *Times* supported it anyway. Soon its advertising audience grew and Science Times became a model for the coverage of science and technology. And the newspaper sold more copies on Tuesdays than on any other weekday.

I have told this story often, because I am proud of the way the *Times* has embraced science coverage. Also, this emphasis on science inspired other papers to embark on similar efforts. By the end of the 1980s, driven by the growth of computer and computer-related advertising, dozens of newspapers had established weekly science and technology pages or sections. The decade also saw the creation and growth of science and engineering magazines.

But this story does not have a happy ending. Today, most of those newspapers have abandoned their science sections, and the vast majority of those that remain focus on health—that is, "news you can use" about products on drugstore shelves, diets, exercise regimens, and so on.

As a result, it is perhaps not surprising that fewer than 10 percent of the 2,400 members of the National Association of Science Writers, established in 1934, are full-time staff reporters or editors for newspapers, popular magazines, or radio and television.[1] Nine percent work for specialty magazines or newsletters. About 40 percent are freelance writers for a variety of publications, and the rest are public information officers for universities, government agencies, private companies, and other organizations, or they are journalism teachers or scholars of journalism.

As news organizations like CNN abolish science reporting teams and the ranks of full-time science reporters shrink, the coverage of science and technology shrinks too. And when science and engineering are covered, often the journalist on the job is someone with little technical background. That is a problem.

Even at the *New York Times,* where some of us (unlike me) trained in technical subjects, we struggle to differentiate the genuinely important discovery from the flash-in-the-pan finding, or to discern which researchers have the support of their colleagues and which are regarded as outliers. And we are a large and (relatively) lavishly supported science news staff. The trouble we have is multiplied for reporters who normally cover politics or zoning boards and suddenly find themselves parachuting into a story about cloning or cold fusion or endangered species or some other technical issue.

Researchers would do well to keep these problems in mind when they are irked by a journalist's performance. But when I talk to researchers about coverage of science and technology, which I do often, journalistic incompetence is one of their perennial themes. I heard the message even more forcefully in 2003, when I wrote an essay for the *Times* on the necessity of having scientists talk to the public. It produced a storm of e-mail from labs all over the place.[2]

Most of the writers started by praising the essay and said I was making an important point. But they did not stop there. A few offered examples of researchers who talked to reporters about their work only to find it wildly hyped in print. "Instead of being able to take simple pride in the article, the scientist is embarrassed his name appears at all," one researcher wrote about a colleague's bad experience with the press.

Another wrote that while "quackulent quotations . . .

are a dime a dozen" the real problem is that journalists do not report the real research enterprise at all. "Very, very rarely are there articles saying 'yet another stunningly complicated and boring experiment confirmed, yet again, something that everyone in the field has accepted for three decades,'" he wrote, yet science advances "not because of one piece of evidence or another, and certainly not because of one news article, but because of an avalanche of evidence; an argument or hypothesis survives a test, and then another, and again, and again. This rarely gets conveyed."

He's right of course. But that kind of thing just isn't "news."

Nancy Baron, an expert on science communication at the Communication Partnership for Science and the Seas in Santa Barbara, California, gets a similar earful when she runs seminars to help scientists learn how to communicate with journalists and to the public. When she asks them what irks them about the news media, some responses emerge again and again:

> Journalists oversell their stories.
>
> News reports do not offer information in enough depth to help readers or viewers form opinions.
>
> Journalists don't see what the important point is and

emphasize side issues that may be more engaging (sexier) but are off point.

Journalists do not understand and therefore cannot accurately explain nonlinear/probabilistic aspects of research.

Journalists confuse science issues with values issues.

Journalists approach their stories with erroneous underlying assumptions.

Journalists look for controversy, or they seek erroneous "balance."

Given all this bile, another item on the list of perennial gripes is rather surprising: "Journalists do not pay enough attention to my field."

These problems are unfortunate and unnecessary, Baron says, because researchers and journalists are a lot alike. We are all curious. We want to find things out, and share the information with others, and we want to do it first. We are analytical. Of every new finding or report we ask, What does this mean? What are its implications? We are critical of our own work and the work of others. And we are highly motivated, persistent, overachieving, independent thinkers who challenge authority, whether the conventional scientific/engineering wisdom or the powers that be ("afflict the comfortable").

But we are a lot different too.

In spite of what many researchers think, science journalists are not advocates for research. At one time, perhaps, we were. At its founding, the National Association of Science Writers had as an explicit goal the advancement of the scientific enterprise.

William L. Laurence, a science writer for the *New York Times,* was seconded to the Manhattan Project and actually rode on one of the chase planes for the bombing of Nagasaki. Even as he covered the beginning of the atomic age for the *Times,* Laurence continued writing press releases for the government and its atomic scientists. In fact, later he was accused of deliberately covering up the incidence of radiation sickness in Japan, at the behest of the U.S. government.

This kind of relationship between a newspaper reporter and a government agency would be unacceptable today (though some people draw uncomfortable parallels with the reporters embedded with military forces in the war in Iraq).

But there are other, more basic differences between researchers and journalists.

Researchers look at things in depth and focus on details. Journalists look for a quick overview. For journalists, details aren't just a nuisance, they can positively interfere with our telling a coherent story. Researchers and journalists often differ markedly in where they draw the line between crucial details and needless clutter.

Perhaps because of their funding environment, researchers emphasize questions, not answers. New findings may elucidate a little more of the known world, but their true practical effect may be to open the door to new questions worthy of new research—and new grants. When, in talking to reporters, researchers emphasize these intriguing uncertainties over the findings themselves, the journalists are likely to imagine the snarl of the prototypical city editor: "Don't tell me what you *don't* know, tell me what you *do* know."

Also, while researchers are rational, journalists are looking for the emotional human element—the frustrations and joys of the research, "the tear-in-the-eye/lump-in-the-throat," as an editor I used to work with puts it. Journalists take this approach to a story not just because it is relatively uncomplicated and easy to tell, but also because it is appealing to readers, viewers, or listeners. If journalists can find a way to tell a story in human terms, they are confident that a lot more of the audience will pay attention to it.

Finally, we tell stories differently. Researchers go from evidence to conclusion. Journalists report the conclusion first, then they put in as much detail as they have room for—often leaving out facts the scientist thinks are crucial. These story-telling variations are behind many of the arguments between scientists and journalists.

Therefore, as Baron says, it is no surprise that researchers view journalists as:

Insufficiently concerned with accuracy

Superficial

Sensationalist

Focused on controversy and tension

Ignorant

Unethical and willing to do anything to get the story

And, it is equally unshocking to find that journalists view researchers as:

Boring

Hair-splitting

Caveating things to death

Overly interested in process

Unable to articulate a bottom line or distinguish the forest for the trees

Users of unintelligible jargon

Are these perceptions the worst problems for science journalism? No—that honor goes to, of all things, journalistic objectivity.

5

The Problem of Objectivity

My 2003 commentary on the necessity of scientists' engaging with journalists was prompted in part by an article by pollster Daniel Yankelovich in *Issues in Science and Technology*, a publication of the National Academy of Sciences.[1] Yankelovich, in turn, was responding to complaints from scientists who, he said, all too often "find themselves pitted in the media against some contrarian, crank or shill who is on hand to provide 'proper balance.'"

"The scientists who hold this view have put their finger on an important problem," I wrote. In striving to be "objective" journalists try to tell all sides of the story.

But it is not always easy for us to tell when a science story really has more than one side—or to know who must be heeded and who can safely be ignored. When we cast too wide a net in search of balance, we can end up painting situations as more complicated or confusing than they really are.

You can think of plenty of examples of such stories: climate change (it's real, and people *are* contributing to it); HIV and AIDS (the virus *does* cause the disease); vaccine preservatives as a cause of autism (*not*). In each case, though, there are prominent people, some with respectable technical credentials, supporting the so-called dissident view. President Bush was a prominent climate-change denier, and it is difficult for a conventional American journalist to say any president is flat-out wrong. One HIV denier was President Thabo Mbeki of South Africa. And the vaccine-autism connection is so embedded in the public mind that it was the subject of the series pilot of a new television show in 2008 (*Eli Stone,* on ABC). When scientists, including the American Academy of Pediatrics, objected to the pilot, the network said its program presented "both sides."

And as I wrote in 2003, "Unless you are an expert, differentiating between the genius and the crank—or even the mainstream and the outlier—may not be easy." For journalists, it may be impossible. This is particularly true because, as a researcher said in an e-mail to me after the article appeared, "the truth value of a scientific statement does not come from the authority of the scientist. It comes from the integrity of the observed data. But I think sometimes journalists forget this."

We don't forget it. We just cannot judge it for ourselves.

As journalists, we are naturally suspicious. I attribute that suspicion in part to our constant exposure to what Stephen Schneider, a climate expert at Stanford, calls "courtroom epistemology," in which people engaged in policy arguments refuse to acknowledge that an issue may have many facets, deliberately leave out inconvenient facts, or present information out of context to advance their position. In courtroom epistemology, no one is obliged to make an opponent's case.

Journalists encounter courtroom epistemology so often we learn to expect it. We may even see it where it does not exist. (Maybe that is why so many people think we are cynical.) So we are suspicious when a scientist presents a one-sided view. We have no way of knowing, on our own, that the evidence is coming down one way, not another. In an effort to give our audience fair and accurate accounts of the news, we may report "both sides of the story," even if one side is much weaker than the other—or even if there hardly is another side at all.

The result, as the environment writer Eugene Linden puts it, is "the systematic over-weighting of dissent."[2]

Lawrence M. Krauss, a physicist at Arizona State University who has spoken out on the teaching of creationism in public schools, among other issues, says there is "an inherent tension" between scientists and journalists when the issue is objectivity. Journalists seek to tell "both sides of the story," he said, "but in science often one side is wrong." For Krauss, journalists' insistence on "finding balance where there is no balance" is infuriating.[3]

As one of Baron's respondents wrote in answer to a questionnaire she distributes in her programs on science communication, "I am most bothered by the concept that 'fair and balanced' is equated with giving equal time to fringe theories. For example, listeners/readers are often left with the impression that a debate exists among experts in a field when in reality the disagreement is being generated by a few vocal and sometimes unqualified outsiders."[4]

We can see this happening all over the landscape of science journalism, but perhaps the most notorious recent example is the coverage of climate change. Climate dissidents (they should not be identified as "skeptics," because skepticism is, or should be, the default position of all researchers) are accorded space beyond their influence in the field. As Naomi Oreskes, a historian of science at the University of California, San Diego, found in her now well known study of almost a thousand research papers on climate change, all accepted the idea that human activity was a big contributor. But most newspaper accounts of the findings included the voices of the "dissidents."[5]

I encountered the problem of objectivity in coverage of the debate over teaching evolution in public schools. There can be intense pressure in the newsroom to give supporters of creationism and its ideological cousins a chance to give their side in what they call the "debate" over evolution. But until someone produces a credible scientific challenge to the theory, there is no debate over evolution. More important, creationism and its cousins rely on supernatural intervention—by definition a hallmark of nonscience. But this is an argument many people do not understand. They think it is only fair to give "the other side" a chance.

It was difficult for journalists to make this point, given that President Bush was on record as believing that "the verdict is still out" on the question of evolution, and especially given that there have been a few people with respectable credentials and affiliations—if not the respect of scientists—speaking out in favor of creationism. As Krauss puts it, "There are a lot of jerks out there, and some of them have PhDs."

I eventually worked out the wording that allows me to sum up the situation, I believe accurately: there is no credible scientific challenge to the theory of evolution as an explanation for the complexity and diversity of life on earth. I use this language often when I write about evolution. But I am criticized for it. On some Web sites, creationists call it "Cornelia's Creed."

The point is, debate and disagreement are hallmarks of science, especially in arenas where science and policy intersect. Journalists will be hard-pressed to tell the good idea from the specious. And when we journalists are unable (or unwilling) to judge for ourselves, we often fall back on he-said-she-said reporting. In effect, we turn the whole question over to our readers (or listeners or viewers), who probably have even less basis for judgment than we do.

In our defense, we journalists know from experience that it is not enough to simply "follow the scientific consensus." Sometimes the consensus is wrong, sometimes spectacularly wrong. I am thinking of the Australian doctors who insisted that stomach ulcers could be caused by infection with a particular microbe, *Helicobacter pylori,* and were roundly ridiculed by mainstream science—until their idea proved so correct it won them a trip to Stockholm and a medal presented by the king of Sweden. Another researcher encountered similar ridicule when he theorized that mad cow disease was caused by microscopic agents he dubbed "prions." He too was scorned, and he too ended up in Stockholm.

When *New York Times* reporter Andrew C. Revkin writes about climate change, he tries to characterize the voices of dissidents as being out of the scientific mainstream. As a practical matter, there is little else that he can do. I worry, though, that the mere mentioning of someone in a news article implies that their views are worth paying attention to.

There are some who say we reporters ought to be able to educate ourselves to make our own scientific judgments. This is the approach Blakemore of ABC took when, after a career as a war correspondent and a Vatican reporter, he suddenly found himself the network's lead reporter on climate change.

Though I admire his energy and diligence, I see a few problems with this approach. First, few news organizations these days will give reporters the time ABC gave Blakemore to master a complicated issue like climate change, especially since there is a tradition of suspicion in journalism about reporters who specialize too narrowly in one beat or another. The fear is they will get too close to their sources. As a result, the more a reporter knows his subject, the more his bosses tend to worry that he is getting too close—and the more likely they are to transfer him to another beat.

Even intelligent and diligent journalists have trouble educating themselves to the point that they can make independent judgments about complex scientific or technical issues. Journalists have been worrying about this problem for years. Universities with journalism programs have expanded their offerings with courses and even whole programs on science coverage. A number of organizations offer summer institutes, semester- or year-long programs, or short "boot camps" on issues in science and technology to help journalists hone their skills. Every year, the Council for the Advancement of Science Writing, the Society of Environmental Journalists, and other groups offer week-long programs in which journalists can learn about cutting-edge developments.

But who participates? Few news organizations will give journalists the time off to attend or cover their travel expenses. (Many journalists use their vacation time and their own funds.) For the semester or year-long programs, many would-be participants worry that their beat—or even their job—won't be there when they return. In an era of journalistic retreat generally and declining research coverage specifically, this worry is well founded.

This problem of objectivity has no ready solution. For me it is by far the most intractable problem in coverage of science and engineering. In my opinion, scientists and engineers are the only ones who can help solve it.

6

The Scientist as Source

In the otherwise forgettable movie *Romantic Comedy,* a character played by Dudley Moore warns a friend about to be interviewed by a reporter: "He's a journalist—he makes a living writing down what people say when they are off guard."

This is a lesson too many researchers have taken too much to heart. They worry so much about saying the wrong thing to journalists that they decide never to talk to them at all. Asked how they deal with reporters, they commonly reply, "I don't take their calls."

This is a mistake. Even if you believe, as many researchers do, that the facts will speak for themselves and make their way inevitably into the public mind, when the facts are complex or unfamiliar they may speak in a language ordinary people (or journalists) cannot understand. They may even be "mute," as Roald Hoffmann, the chemist and Nobel laureate, wrote in an essay in *American Scientist,* generating "neither the desire to understand nor appeals for the patronage that science requires, nor the judgment to do A instead of B, nor the will to overcome a seemingly insuperable failure."[1]

In an ideal world, you would convey your facts in a way that inspires that kind of thinking. At a minimum, you want to convey your facts in a way that allows no misinterpretation. Is this perfection easy to achieve? No. But you can greatly improve your odds of success.

The first step is to realize why so many scientists have so much trouble with so many reporters: lack of preparation, on both sides. Speaking as a journalist, I confess up front that lack of adequate preparation is always going to be a problem for us. Even if we work at news organizations with relatively abundant resources, we can never be as up to speed as we would like to be about all the developments in the areas we cover. At a news organization with only one science writer—or no full-time science writer—reporters will be chronically unprepared. Researchers need to understand this fact of life and take steps to cope.

WHEN A REPORTER CALLS

So, let's imagine that a reporter has called, with questions about your work or someone else's work that is in the news. Here are some simple steps to take.

Ask what in particular the reporter is calling about. (Be aware that the focus of a story can shift, if further reporting suggests that it ought to.) Ask what the deadline is and when the story will run. Someone who has an hour to file a spot news story is going to want a lot less information than someone who is "collecting string" for a project to run in the indefinite future.

Now that you know what the reporter is calling about and how much time she has, ask for time to collect your own thoughts. That is, tell the reporter you will call her back—in five minutes if time is short, or in an hour, a day, a few days. Insist on at least five minutes. Even a reporter on a red-hot deadline can spare you five minutes. But do not make this offer unless you absolutely intend to call back at the agreed-on time.

If you have time, google the reporter. Look at other articles she has written and see whether you think she did a good job. Did she make mistakes? Exhibit common misinformation? Be ready to correct her. Perhaps you will discover something that will make you decide she is not someone you want to talk to at all. I hope not. But if you do decide not to pursue the interview, call back and tell her you will not be able to speak with her. (You don't have to go into detail.)

If you agree to be interviewed, imagine you are approaching your own field from a position of ignorance. What would you need to know? How much could you absorb in the short time it takes to read a newspaper article or watch a televised news segment? Figure out the main point you want to convey—or the two main points or, *absolutely tops,* three main points. Figure out how to make these points in the clearest, simplest language you can—retaining crucial details but ditching irrelevant clutter. (Be ruthless.)

In their book *Speaking about Science,* Scott Morgan and Barrett Whitener talk about "the money slide," the slide every scientist should show while making a presentation to fellow researchers.[2] This slide should encompass the most important finding, or draw the data together into a coherent whole. It should be easy to describe and easy to understand. If you were giving a professional talk, you would value this slide. Prepare the equivalent when you have to talk to a reporter. It is your message.

Can you complete this preparation in five minutes?

Probably not. That means you must be thinking about your message even before a journalist expresses any interest in your work. If you are smart, you will do this work way ahead of time. In fact, whenever a friend or family member asks you what you're up to in the lab, consider it an opportunity to perfect your message. Practice won't necessarily make you perfect and preparation won't necessarily turn you into a media star but they can help you speak much more easily and clearly.

"You have to go into an interview with something to say," my fellow presenter, Frank Kauffman of Edelman Public Relations, told participants in June 2008 at the Aldo Leopold Leadership Program session in West Cornwall, Connecticut. "You cannot just be in response mode."

Kauffman, who was a colleague of mine at the *Providence Journal* and later a reporter at the *Baltimore Sun,* views interviews as opportunities to deliver a message. And, he says, "a message is not a fact, a message is a point of view. Facts, like statistics, prop up the message, but the message is bigger than a fact."

So take the time to figure out what your message ought to be, and then think about how to convey it. Imagine how you would describe the things you work with if you could not show a photo or provide a drawing: "It looks like . . . moves like . . . appears as if . . . acts like . . ." Relevant analogies or apt metaphors will not only help the reporter understand you but will also help her explain you to her audience.

If you have ever taught, you have probably said something in a lecture or seminar that your listeners did not get. You saw the veil of incomprehension drop before their eyes. While you are talking with the reporter, look and listen for hints that something similar is happening with her. This kind of observation is obviously easier in person than in a telephone interview, but even in a phone conversation there are clues. Listen to the reporter's questions and comments. Are they appropriate? Pay close attention to whether the journalist is following you.

Also, try to ascertain at the outset how much information the reporter is bringing to the interview—and how much of it is wrong. If there is a common misconception about your field of research or a mistake people often make, point it out explicitly. Be prepared to go where the interview takes you, but beware of wandering down irrelevant byways—and be especially aware that the journalist may not recognize them as such. She may not know the difference between what is central and what is peripheral unless you tell her.

If the situation is not black and white, say that explicitly and emphasize the point. At the same time, consider whether a situation that seems gray to you may be black-and-white enough for a newspaper or television audience.

When talking about numbers, say what they are but also what they mean. Give an example and say what it signifies. Percentages and ratios are easier for most people to understand on the fly than absolute numbers, but they can be deceptive. For example, saying that a chemical in the environment doubles a person's risk of contracting a particular kind of cancer is a lot scarier than saying it raises the odds from one in ten million to two in ten million. That kind of doubling is very different from a doubling that raises the risk from three in ten to six in ten. Think about this problem ahead of time and work out the clearest and most accurate way to present your data. Maybe you want to prepare a chart or graph for the reporter.

In fact, spend some time considering how the journalist could illustrate your work—on paper, on the air, or online. As the National Science Foundation and the American Association for the Advancement of Science put it in materials about their award for efforts to visualize information, "Science and engineering's most powerful statements are not made from words alone." Think about the photos, maps, videos, charts, graphs, sketches, or other "art" you can provide. As Steve Duenes, graphics director for the *New York Times* (and the graphics editor in science when I was science editor), suggested in an online discussion, identify the "interesting or important patterns in the data" so the news agency can display them.[3] Give or send this material to the reporter—if possible, before your interview begins.

If English is not your native language and you are going to be interviewed in English, spend a little time talking with a friend or colleague who is a native speaker and is willing to correct any mistakes you might make. (Native speakers ought to do this too but don't.)

Once you are talking to the journalist, think about techniques Frank Kauffman calls bridging, flagging, and repetition.

Bridging is what you do when you are asked a question that does not relate to what you want to talk about. If that happens, don't blow the question off, but look for ways to "drive to your message," as Kauffman puts it, with phrases like, "Yes, and in addition" or "Yes, but the real issue is" or "What's important to remember . . ." (Of course, if the question can be answered in a word or two, just answer it. "Where do you teach?" "Worcester Polytech.")

You can use bridging when you are asked about something outside your area of expertise but you have something else you want to contribute. Here you can say something like, "I don't know about that, but what I *do* know . . ."

If the question is something you would really rather not talk about, answer it as much as you can and then move on. Kauffman cites former president Bill Clinton as a master of this technique. When he was running for president in 1992 he was asked on the news program *60 Minutes* if he had had an affair with a woman in Arkansas. He answered by admitting he had caused "pain" in his marriage and then bridged to issues he said Americans were far more worried about. In other words, he answered the question without blowing it off, and then took the conversation where he wanted it to go.

Flagging is simply making sure the reporter knows what the important points are—which ones he should underline or mark with a star in his notebook. Flag important points with language like "The most important point is" or "If you remember nothing else . . ." This kind of guidance can be immensely valuable to a reporter.

Finally, repeat your message. "Once is not enough," Kauffman told the Aldo Leopold Leadership Program. "Once is not enough."

Some other tips:

Don't interrupt the questioner. (This guidance ought to be unnecessary, but my own experience as a journalist tells me it is not.) Kauffman advises counting silently to two after the question is asked before you start to speak.

When you have made your point, stop talking.

Ask how to reach the reporter, in the relevant time frame, in case you remember something you ought to have said or think of something you need to clarify. Tell the reporter how to reach you. (She should ask, but if she doesn't, offer the information.)

Remember that the reporter is not your friend. If she calls or e-mails you with questions, it is probably because she wants to quote you. Don't reply as you might to a friend or colleague—a temptation if the journalist is someone who has covered your field for a while and is someone you have gotten to know.

I once e-mailed a scientist I knew to get his opinion about a piece of engineering that was touted as a way of getting carbon-free power from the geothermal energy of the earth. He e-mailed back: "I know all about it and it's mostly bullshit." A few minutes later I received another e-mail from him. This message said, "If you are going to quote me, I will send you an e-mail with a more scholarly answer to your question." It is best to offer a measured response in the first place.

For the reporter, talking to you is a business activity. Don't take anything personally. Don't get angry. Don't be impatient with questions you think are stupid. Remember that the reporter cannot possibly have the knowledge you have.

Don't think of preparing for these encounters as make-work. Add it to your to-do list—and then move it toward the top. Make this task a priority. You won't necessarily turn yourself into a media star, but you may become a useful source. The actor Alan Alda, who worked with many scientists in the television series *Scientific American Frontiers,* draws the analogy to method acting. Such training cannot create emotive genius all by itself, he once told me, but "it has turned many an untalented person into an adequate performer."[4]

THE SOUND BITE IS YOUR FRIEND

When a journalist calls, she is looking for terse, telling quotes that sum things up in a way that an ordinary person will understand. In other words, she needs sound bites.

For many researchers, this need proves all by itself that journalism is a superficial, pointless enterprise doomed to confuse or even deceive the public. But there will be sound bites. If journalists do not get them from you, they will find them elsewhere or compose their own. Will their sound bites be as accurate and cogent as yours? Of course not. So be ready with your own.

The Centers for Disease Control and Prevention

made this point when it issued "Guidelines for Investigating Clusters of Health Events" in 1990. A cluster of health events—a cancer cluster, say, or an unusually high number of children with autism in a given school—is almost never a sign of anything sinister or even dangerous. But, as we have seen, people won't necessarily take this sanguine view. They will be worried, and vulnerable to misleading information. And so, the CDC advised,

> investigators must realize that the media tend to simplify complex, technical explanations, thereby losing subtle distinctions or qualifications. Thus, investigators should distill the messages they wish to convey and present them in the way they are most likely to be transmitted without confusion or distortion. Investigators must be prepared to stress key points; provide background necessary for understanding; and be straightforward regarding what is fact, what is speculation and what is not known. Most of all, investigators must remain cooperative and responsive and must be prepared to provide needed information rapidly, before distortion and discord have been introduced into public exchanges.[5]

This is good advice, even if you are not at the center of a public health emergency.

I learned the power of the sound bite when I wrote a book about beach erosion and coastal land use. When people asked me what my book was about, I would drone on about sediment transport, sea-level rise, building permits, repeat flood insurance claims, and so on. Finally Sheryl Gay Stolberg, a colleague at the *Times,* pointed out that, while I was clearly fascinated by my subject, I was boring everyone who came near me. She said, "Tell them, 'Through greed and ignorance, Americans are destroying the landscape they love the best, the beach.'"

That was my sound bite and I have used it ever since. And I have developed others for particular segments of my topic. They are accurate (or accurate enough), pithy, understandable, and engaging. You should have similar formulations for your work. Figure out ahead of time what they are and be ready with them. Once you have one in mind, ask yourself if you would be happy to see it in a newspaper headline. If the answer is yes, you are onto something.

Take a lesson from the Republican pollster Frank Luntz. Among other things, Luntz was an engineer of the party's efforts to block government action on climate change, in part by shaping the way the issue was discussed in the political arena. Even if you deplore his actions, you have to admit his approach was effective.

Luntz is the master of sound bites—what they should be, their importance as a genre, and what happens when news sources don't have them.

According to Luntz, Rudolph Giuliani's failure to master the sound bite was what drove the former mayor of New York City out of the 2008 presidential race: he never provided "crisp, clear, concise sound bites."[6] Instead, Luntz said, the former mayor indulged "in long-winded answers clogged with detailed statistics. Audiences were uninspired and, sometimes, baffled."

Does that sound like you?

Unless you are a natural media personality with perfect pitch and on-the-spot glibness, useful sound bites will not come to you by magic when a journalist calls. You will have to prepare them.

Think ahead of time about whether other people—the journalist and her lay audience—will know what you mean by terms like *genotype* or *isobar* or *muon*. Avoid jargon, acronyms, abbreviations, and other terms someone outside your field should not be expected to understand. When in doubt, err on the side of explaining the term or substituting a lay-language equivalent.

We ponder this kind of question a lot at the *Times*, and we ask ourselves whether the term in question is "a headline word." That is, if it appeared in a headline would readers know what it meant? For example, for our readers we think "DNA" is a headline word. But RNA is not, or not yet anyway.

Use the same thought process when you want to build analogies around cultural references, because it is no longer reasonable to assume that the generally educated person will know what you are talking about. This unfortunate development was brought home to me when I watched an episode of *Monty Python's Personal Best,* in which members of the BBC's beyond-loony comedy troupe discussed the sketches they liked the best.

Among John Cleese's favorites, he said, was a supposed news broadcast of a competition among painters making art while bicycling along British motorways. Cleese played the commentator, offering updates from a particular traffic circle on the work of one contestant, Pablo Picasso. The bit concluded with his rundown of other contestants, including Raoul Dufy and Georges Braque. Finally, a little person dressed in black and wearing a slouch hat (Henri de Toulouse-Lautrec) rode by on a tricycle.

The sketch dated from the 1980s. "We could not do it today," Cleese said sadly—because today most people don't know enough about art to get the joke. You may regret this fact of life. I do. But you must consider it.

Otherwise you will fire allusions over the heads of your audience.

Also, if you use literary or other allusions, or foreign words or phrases, you should be absolutely certain that *you* have them right, not just in terms of their grammar, spelling, punctuation, and so on but also their larger meaning. Actually, it is probably best to avoid such fancy touches in the first place. (According to the magazine *The Week,* a British furniture chain got into trouble on this score when it began marketing a bed designed for little girls. Its model name: Lolita. One can only hope the marketers had not read Nabokov's novel.)

Now ask yourself whether I, in the previous paragraph, assume too much about you.

TALKING ABOUT A STUDY

Many scientists first hear from a reporter when they report something newsworthy in the scientific literature. When you receive this kind of call, make sure the journalist has a copy of the paper and the editorial, commentary, or counterargument that accompanied it, if any. Then figure out how to restate, in lay language, what you say in your report. Start with your main point: what is the meaning or significance of this work? How does it move the ball in your field?

Point out the paper's description of the study's original purpose, research design, methods of data collection and analysis, number of subjects, who or what they are (people, animals, what kind, and so on). If the study does not involve people, say what its implications for people are, if any. Similarly, if the finding has implications for public policy, clinical standards, or the like, say why. If there are risk/benefit trade-offs involved, describe them too.

If there is something novel about the design or methods of your research, say so. Sometimes one of the most interesting things about a study is the way it was conducted, but the journalist may not realize this.

Acknowledge the limitations or shortcomings of the work. If findings are preliminary or inconclusive, make that clear. If, on the other hand, the results are surprising—if they challenge previous work, for example—say so and say why. And provide some evidence.

If you want to be truly helpful, tell the journalist who else could comment on your work, especially if they are unconnected to it or (even better, from a journalist's perspective) you respect them but they are likely to disagree with your conclusions. Say who financed your work, what stake they have in its outcome, and what influence or control they had over the report itself.

If you have submitted a paper to a journal and it has not yet been published, be familiar with the journal's policies on when you may talk about it with the public (journalists). Some will let you speak with reporters before the article is published, but only if reporters agree not to publish anything before the publication date of the issue in which the report will appear. Some journals penalize scientists who let the world know about their work before it appears in print—they may even decline to publish a paper they had previously accepted. This practice is deplorable, but it is a fact of life in science, so protect yourself.

You can talk to reporters on condition that they respect the publication's embargo policy. In other words, tell them you'll speak, but that they cannot publish your comments until the time set by the journal. Be sure to establish these ground rules *in advance.* And don't talk to a reporter on this basis unless you trust her to keep her word.

Lots of people object to embargo policies on principle. They say embargoes encourage a herd mentality among reporters and can result in a torrent of breathless reports on small-bore findings. Others say embargoes discourage "enterprise" reporting in favor of the "spot news" of a journal report. I don't understand those arguments. Reporters should be able to make coverage decisions about journal reports regardless of whether they are covered by an embargo.

OFF THE RECORD

If you are talking to a reporter, you should assume everything you say is being written down or recorded and might appear in print, online, or on the air. That is, assume you are "on the record." After all, the journalist is talking to you in the hope of obtaining information she can use in her report. You should not say anything you would be unhappy to see or hear attributed to you in public.

There are practical reasons for this advice. For one thing, there is no rule book for journalists that defines conditions like "off the record" or related states like "on background," "on deep background," "not for attribution," and other terms of journalistic art describing the quotability of what you may say.

For example, I believe a journalist who obtains information not for attribution or on background may use it but may not identify its source, or must describe the source only vaguely—as, say, "a biochemist familiar with the new work" or "a mechanical engineer who has followed the dispute." A journalist who obtains information "on deep background" may use it in her stories but may not attribute it to anyone.

Other journalists may have different ideas. If you want to speak off the record, establish the ground rules clearly *before* you start to talk. Be specific. And spell out how much detail the journalist may use to describe you if you don't want to be named.

But before you get that far, ask yourself whether you can rely on the reporter to keep your identity secret. Though there continues to be agitation for a so-called journalists' shield law, a law that would grant conversations with reporters something like the protection given conversations with lawyers or pastors, such a law does not yet exist nationally, and state laws vary in their protections and the degree to which they have been litigated, which is to say, the degree to which their meaning has been legally established. In the most telling Supreme Court case on the issue, *Branzburg v. Hayes,* the Court ruled in 1972 that unless reporters could demonstrate that they were the victims of prosecutorial abuse, they could be compelled to testify.[7]

As a journalist, I believe there are many reasons to protect reporters from being compelled to testify. It is not necessary to go into them here. It is enough to say that, in general, the protection does not yet exist. So if you are tempted to speak off the record, and if the issue is or could become controversial, ask yourself if you know the reporter well enough to be confident she would go to jail rather than give you up.

The use of quotes attributed to anonymous sources is a problem for news organizations anyway. Information attributed to someone who is identified by name and affiliation carries more credibility than information attributed to an anonymous source. Also, attributing information to anonymous sources can allow unscrupulous reporters to fabricate material. That may be one reason why in recent years there have been somewhat fewer stories published that rely on anonymous sources, and why, when a source is granted anonymity, the reason is usually made plain.

Of course, in the annals of journalism there are many examples of important stories that saw the light of day only because people spoke—and journalists were willing to quote them—off the record. Perhaps Watergate is the most famous of the stories and "Deep Throat," so called because he spoke on deep background, is the most famous source. Mark Felt, a retired FBI official, finally outed himself as Deep Throat in 2005, about thirty years after the stories appeared. Carl Bernstein, one of the *Washington Post* reporters who broke the Watergate story, has often said that when he and Bob Woodward were reporting on Watergate, the only people who consistently told them the truth were the people who spoke off the record.

PEER REVIEW

There are some scientists who believe they should be able to review a journalist's story before it appears, to check for inaccuracies. They say they consider it a kind of peer review, normal in science. But it is a controversial issue in journalism.

Though I will often call researchers back to ask them to confirm my understanding of what they have told me, I have never acceded to this kind of "peer review" request, on the ground that I would not let the mayor or the police chief or a corporate leader or anyone else read any story of mine before it was published. Also, I believe scientists and engineers are as prone to rivalries, egomania, and ulterior motives as anyone else, and I don't want to let them into my writing process any more than necessary.

Other journalists don't hold this view, however, so if you feel strongly, it is worth asking about. Just don't be surprised if you are turned down. Then you have to decide whether you will agree to be interviewed anyway.

If you are really nervous about being misquoted and time is not pressing, ask the journalist to e-mail you questions for you to answer by return e-mail. I consider this a last resort when I am interviewing someone, because, except in really cut-and-dried issues, conversations are usually more fruitful—and they help me produce much better, livelier copy. But an e-mail exchange is better than nothing.

CORRECTIONS

According to the Union of Concerned Scientists' media guide for scientists, errors are the price many scientists think they have to pay for coverage of their work.[8] I am not sure I agree with that—it should be possible for journalists to get things right. Scientists certainly should not assume stories they are interviewed for will inevitably include errors.

Nevertheless, everyone makes mistakes and journalists' are out there for the world to see or hear—and repeat. However mortifying it is to admit we have made an error, it is important for us to correct them. And it is deeply frustrating to hear a researcher say it is pointless to call a reporter to report an error because journalists don't care about accuracy. We care passionately about accuracy. It benefits us not at all to be associated with a story, even if it is big news, if it has errors in it.

Also, nowadays news stories live on in electronic databases. So do the errors they contain. Unless you tell us when we make a mistake so we can note it in the database, anyone who consults the stored version of the article in the future may repeat the error.

So if coverage is botched, let the reporter know. Call and say, "Thank you for your interest, but there is a problem with the story." Then tell the reporter how to correct it.

Keep in mind that there is a difference, as Sharon Dunwoody of the University of Wisconsin School of Journalism and Mass Communication puts it, between "technical accuracy" and "communicative accuracy." She is talking about the difference between omitting the name of a study's coauthor, say, and giving an erroneous explanation for a scientific finding. In other words, does the error cause people to misunderstand something about the article? In that case, it needs a correction. Is it a trivial omission? Then maybe it does not.

At the *Times,* we address this issue by asking if an error "rises to the level of a correction," and in general we err on the side of running them.

Years ago, the *Times* developed a formula for writing corrections. Below is a version of that formula you can use when you think a correction may be warranted. If the issue you are griped about can be expressed in this format, suggest the correction.

1. Begin with a phrase describing the article at issue, when it appeared and, if useful, where.
2. Describe what the article was about—enough so people who read your correction will know what the subject was.
3. Say what the error was.
4. Correct the error.

For example, you might write: An article in Science Times Tuesday about the teaching of evolution referred incorrectly to the number of states in the Union. There are fifty, not forty-eight.

At the *Times,* corrections like these are often amplified with explanations of how the error occurred (an editing error, a transmission error, and so on) and when it appeared (in some editions, in early editions). But you don't need to worry about that kind of thing.

We also had a formula for editors' notes. Here's a version you can use:

1. Begin with the same description of the article and when/where it appeared.
2. Describe what it was about.
3. Describe the matter that is griping you.
4. Say what you think, in fairness, ought to have been included/omitted/changed.

For example, you might write: An article in Science Times Tuesday about the teaching of evolution reported an accusation by school officials in Somewhere, New Jersey, that Mary Jones, a mathematics teacher at Somewhere High School, teaches creationist ideas in her classes. Ms. Jones does no such thing, and the accusation should not have been allowed to stand unchallenged.

The editors might alter your formulation to say that Ms. Jones says she does no such thing and to add that it is the policy of the newspaper to allow targets of criticism to have their say. They might also add that her denial was inadvertently dropped because of an editing lapse (or whatever). Nevertheless, if you can express your complaint according to this formula, you are on to something.

Think about how you would write a correction, following these guidelines. If you cannot do that, think about whether the mistake you spotted is an actual error or merely an omission of a noncrucial detail or something so trivial that a correction would end up as a parody of itself.

Also, if you know a reporter is mistaken about something, don't wait until she puts it on the air or in a newspaper before correcting it. Years ago, at a conference, I met a prominent marine ecologist. I got the spelling of his name wrong, wrote it incorrectly in the directory of news sources I keep, and have been misspelling his name ever since. I discovered my error only when I wrote about him in the newspaper.

A WORD OF THANKS

A journalist working on a complicated story will talk to many sources. If you are one of them, you may speak for an hour and end up with one sentence in the article or on the air. Maybe you won't be mentioned at all. Please do not assume that the time you spent talking to the reporter was wasted. You imparted information that contributed to the reporter's understanding of the issue and, as a result, the quality of the overall report. You left the journalist better informed for her next article on the subject. You performed a public service.

When Kenneth R. Weiss and Usha Lee McFarling received the Walter Sullivan Award for Excellence in Science Journalism from the American Geophysical Union for "Altered Oceans," their exhaustive survey of the state of the oceans, a project for the *Los Angeles Times* that had already won every other major journalism prize, including the Pulitzer, they went out of their way to thank the people who had ended up on the cutting room floor. "We extend an extra appreciation to the many scientists who generously gave us their time and insights, even though their names did not appear in the articles," they wrote. "These unnamed contributors added immeasurably to the articles' depth and breadth."[9]

ANOTHER WORD OF THANKS

Finally, if a reporter does a good job, call or e-mail to say so. You have no idea how much that means to journalists, especially when they are struggling to do a good job with complex material.

7

Public Relations

According to Earle M. Holland, assistant vice president for research communications at Ohio State University, most researchers have "an infantile belief" that scientific findings routinely make their way to the public without human intervention. As he put it, "They envision a 'eureka moment' on the part of a researcher, after which that new knowledge osmotically wafts from the laboratory into the populace." While this idea is "comforting," he wrote, "I've never seen it happen in real life."[1]

The problem, Holland says, is that most researchers are not students of the media. They don't know how news organizations operate and they don't have practice dealing with them. Holland does not expect them to. He expects them to turn to the people most universities or scientific institutions employ to navigate these shoaly waters. They may be in the organization's office of community relations or its government affairs bureau or its news office. What they do is public relations.

Some public relations specialists may have trained as science writers, or even as scientists. Some may be former journalists with no special technical knowledge. But what they should have is knowledge of how the media work and how best to approach journalists with information the public might want or need to have. As with any human endeavor (including journalism *and* research), practitioners of science PR vary widely. Practitioners like Holland can help you a lot.

Again, preparation is key. Find out who does this work at your institution. Ask your colleagues about them. Don't wait until a reporter is on the phone to learn who at your institution deals with the media. And when you seek guidance from a public relations specialist, pay attention to what he says. You don't necessarily have to accept every single bit of advice he offers, but consider all of it.

Be aware that your institution may have policies on when or even whether employees may talk with the media. Naturally, as a journalist I think everyone should be allowed to talk freely with reporters. But if you are going to be in trouble unless you go through your news office or public affairs people, you had better know this ahead of time and make your decisions accordingly.

A good public information officer can help you in many ways. He will know what publications, broadcast outlets, and other news organizations are interested in your field and who reports on it. He will be on top of what's new. Dennis Meredith, an expert practitioner who spent decades in science public relations for such universities as Cornell and Duke, noted that many people in the PR field attend scientific and technical conferences in hopes of corralling journalists and persuading them to write about their institutions. That was never Meredith's plan. "I don't go to pitch stories," Meredith told me.[2] "It's more to find out who's writing about what." For public information officers, or PIOs, this kind of knowledge is invaluable.

A good PIO has news sense—he will know, say, whether a particular laboratory finding is newsworthy and for what audience. He will warn you not to expect breathless coverage of everything and anything your lab might produce.

According to Rick Borchelt, who has worked as a PIO for Congress, the Oak Ridge National Laboratory in Tennessee, and the Broad Institute in Cambridge,

Massachusetts, and who is a founding member of the Society of Environmental Journalists, the job of the public information officer is to nurture "engagement" between scientists and the public. This engagement is not the mere transmittal of a series of facts. "It should be a matter of trust," he said.[3]

Borchelt says you can assume that "you have credibility a priori if a journalist is calling you. But you can lose it," he warns, by making too many claims about your own work, or by ignoring the work of competitors or critics. A competent public information officer understands that successful relationships between researchers and the public are not a matter of one-time-only flashes of scientific publicity. They take time to build.

THE PRESS RELEASE

A basic tool in this effort is the press release, a notice sent out in the mail or by e-mail (or even on a videotape or CD or a posting on Facebook or YouTube) that tells journalists about a research finding or other news. The first thing to know about press releases is this: they should be limited to real news. A good public relations person can tell you what that is. What isn't news, except in your hometown, and maybe then only if it's a small hometown, is a faculty appointment (unless the faculty member in question is a celebrity), a promotion, a routine award or grant, a new facility (unless it is truly unusual), or a finding that may be valuable but does not significantly move the ball on your particular playing field.

A good PR person knows not to blanket the earth with announcements of this nature, even though new media technologies make it easy and inexpensive to do so. He wants to be paid attention to.

Meredith used to say that he avoided the mass mailings and flashy gimmicks beloved by some public relations people because he wanted reporters to know that when they heard from him, it was because he had something interesting to say. Speaking as a journalist who regularly received his mail, I can testify that the tactic works. Borchelt and Holland and others who follow this practice win the respect of journalists, who come to rely on them. Researchers who have colleagues like them are fortunate.

"Establish your credibility by putting out news releases that are real news," Meredith told me years ago, when I asked him about his success. "My philosophy is, every press release has to be good enough that a media person thinks it's good enough to use, even if they don't use it."[4]

Releases about new findings and other hard news should be played straight, he said—"very tersely written, you don't get cute, you don't back into the lede" (the lede—a journalism term—being the most important thought, which ordinarily appears at the beginning or "top" of the article). At the same time, he went on, "The press release has to satisfy every stripe of science writer who will read it. They have to be extensive enough so that the *New York Times* can pull out what it needs, but also *Mechanical Engineering News.*" Detailed information may not find its way into print or on the air, Meredith said, but it increases the credibility of the news release. "It convinces them you have a handle on the story."

And of course, he said, "you should always include who else is doing similar research at other institutions. A lot of people won't do that because they feel it's advertising for the competition, but in a real sense it helps." He is right. And that means, as he put it, "we must keep up with what's going on elsewhere." That's something you can help with if a public relations person is writing a news release about your work.

Meredith points out that while it can be tempting to announce research findings when they are made, it may be advisable to wait until they are published in the scholarly literature. "You don't want to put it out to the lay media until there are vetted results," he said. (Often people criticize science writers for waiting to write about a finding until it has been reported in a scholarly publication, but I think such caution can be crucial.)

When it comes to the release itself, my former *New York Times* colleague Warren Leary says he is grateful when scientists and public relations people take a two-part approach, issuing at the same time what he calls a "rip-and-read" summary of whatever is being announced or reported and a longer, more detailed account including links to other sources of information, possible photos, or other art, and so on.

I agree. A short summary allows the journalist to assess quickly whether he needs to take the time to look into whatever this news is. The longer piece gets him started.

A press release should include contact information for the public relations people *and* for the researchers who can talk about the work. If you know you have been listed on a press release that is about to go out, make sure you can be reached at the phone or e-mail contacts listed.

The release should tell reporters—in an obvious way—if the information is covered by an embargo—that is, whether it must not be made public before a particular date and time. Journals with embargoes may issue press releases about the studies they publish and require journalists who receive them not to report the findings until the publication date of the journal. If you do not include embargo information in a press release, journalists will assume they are free to use the information as soon as they receive it.

If possible, include maps, charts, photos, drawings, and any other helpful art. At a minimum, provide a link to a site where this material can be found. If they are relevant, include fact sheets. John Holdren, a climate expert at Harvard and President Obama's science adviser, used to accompany materials about his work on climate change with fact sheets including temperature data and so on. This kind of information can be extremely helpful to journalists.

If your release relates to the publication of a study, include a copy of the paper or at least a link to it. You don't need to provide full-fledged biographies of all the researchers involved, but it can help journalists if you offer a little information about who did the work and (especially) how to reach them.

Be sure you will have a chance to review any press release about your work before it goes out to the world. It is not uncommon for scientists to be pressed by public relations offices to make more of their findings than the findings themselves warrant. Universities, eyeing the *U.S. News* rankings and impressionable potential donors, are eager to bang their own drums. If your work was financed by a private company, that company may seek commercial advancement in overplaying your results. Some journals too, possibly because they are commercial operations, are eager to portray themselves as publishing high-impact work, and may hype your paper in a press release in advance of publication.

Don't be pressured to make more of your findings than they are worth. Dealing with this pressure at the outset, no matter how unpleasant, is easier than having to deal with a PR overshoot after it has occurred. Again, thinking *ahead of time* about your work and how to describe it accurately will help you here. You might be able to defuse the situation by saying something like, "What you propose to say in your press release goes beyond what our data can support, but what we *can* say is . . ."

Even if your PIO is highly competent and high-minded, the press release "is sometimes a minefield for both parties," Holland notes, because the scientist and the public information officer may have different ideas about how much it should contain, what level of detail is important, and how precise it needs to be.[5] Sometimes coming to agreement on a news release can be a complicated negotiation.

When you find yourself in such a situation, keep in mind that while you want the release to be accurate (and the PIO does too), the public information officer also wants it to draw the attention of the public. He does not want it to be so filled with trivial details, caveats, and irrelevancies that no one will pay any attention to it. You both will have to settle on a release that is accurate "enough."

This process takes time. If you have a paper coming out that you think will draw public attention, or ought to, give your public information colleague as much notice as you can. And if you find yourself in the middle of a hot breaking news story, you will have to be prepared to drop what you are doing and turn to the release the moment it is ready. Do not neglect this task. Even a modest amount of attention paid before a release goes out can forestall major trouble.

THE PRESS CONFERENCE

You should think twice about attempting to hold a press conference without the help of a public information officer or other person designated to run interference between you and the media.

Among the questions the PIO can help you answer ahead of time are: who should appear at the conference, who should be on hand to offer supplemental information or moral support, who will speak and for how long. He can also advise you on which reporters to invite. But be aware that if you invite reporter X, his news organization may send reporter Y instead. Realize, too, that others not specifically invited may hear about your press conference and turn up for it.

The PIO can also help arrange for the needs of the electronic reporting media—adequate electrical outlets, phone hookups, and so on. Remember that nowadays every journalist will have a computer and many who work for print outlets will be expected to provide video or audio with their reports.

Begin the press conference by thanking the journalists for turning up and then open with a prepared statement or statements, keeping things brief and to the point. Be disciplined about this. Opening statements of more than five minutes or so are probably too long, especially if several people are making opening statements.

After the statements, take questions, one at a time. Do not allow any one reporter or news outlet to dominate. Your public relations person can act as moderator.

If someone asks a question calling for a large amount of technical material or the like, say you'll talk later with him or her or will follow up with additional material. Be prepared for questions like this, and be prepared to deliver at the time or soon after—perhaps with fact sheets or reprints of research reports. If a reporter challenges your information with data you do not know about, don't accept his assertions until you're sure he's right. Tell him you will get back to him.

Have a time limit in mind (but be flexible) and a procedure for closing the proceedings. (The PIO can say "we have time for one more question," for example.)

WHEN THE NEWS IS BAD

With luck, you will never find yourself in the middle of a "bad news" story, but it can happen to the most hard-working, honorable person. Misconduct and fraud are obvious problems, but, as Holland notes, research involving human subjects, laboratory animals, radiation safety, biohazards, computer security, privacy issues, infection control, and a host of other issues are inherently "troublesome realms."[6]

If you find yourself caught up in what practitioners call "crisis communication," a good public information officer can be immensely helpful. As he will tell you, the two chief rules for dealing with a crisis are: tell the truth, and tell it quickly.

Often institutions are reluctant to stick by these rules, though, because they want to hush things up, for fear the public will react negatively. But, as has been amply demonstrated in crisis after crisis over the years, the public is much better able to deal with issues that are openly acknowledged. And in the YouTube era, it is no longer possible to sit tight and hope something ugly will simply go away. As Holland writes, "If there is something gone awry that is awaiting discovery, then organizations and individuals are running out of time as they wait."[7] People will be hearing about the situation. Make sure they hear your views.

8

Telling Stories on Radio and TV

The writer Gore Vidal is said to have embraced this simple rule of life: "Never pass up an opportunity to have sex or appear on television." You may not think much of this adage—the television part, anyway. So you may not think researchers need to be adept at telling stories on television or radio. You may even fall into the notorious researcher's trap of thinking that those who speak smoothly and engagingly on the air have embraced superficiality or, worse, have somehow betrayed their research for the sake of fleeting fame.

But some of the best science reporting today is done on public radio and public television, and there are outposts of excellence in commercial broadcasting as well. Though individual broadcast and cable television network audiences have shrunk, overall the audience is still huge, numbering in the scores of millions.

And just as radio did not kill newspapers, movies did not kill radio, and television did not kill movies, the advent of the Internet is not going to kill the mainstream media per se. It is going to spread the techniques of these media in new, more diverse directions.

Plus—and this is the most compelling argument—as newspaper Web sites claim more and more newspaper readers, and broadcast outlets' Web sites become more and more central to their operations, journalists of all kinds are increasingly using the tools of television and radio to tell stories on the Web. So you ignore broadcasting skills at your peril. You must understand how to use the electronic media effectively.

The first thing to understand about the broadcast media is their time frame. Jeff Burnside, an Emmy-winning reporter at NBC in Miami who regularly speaks to researchers about the media, tells of approaching a marine biologist about a possible television feature on his work. Sure, the scientist said, so long as you can give me twenty minutes of airtime to explain it. Not an unreasonable request—unless you know that, after commercials and promos, the average half-hour newscast has only about eighteen minutes for all of its stories. This scientist was asking for 110 percent of the entire broadcast.

The average item on a typical commercial television news broadcast lasts twenty seconds or less, sometimes much less. In this environment, a two-minute feature is a big investment of television time—about the most you can reasonably expect on a news broadcast. All of which is another way of saying that figuring out what your most important points are, and how to express them succinctly, is absolutely crucial on television.

As the writer Michael Eric Dyson put it in an interview with the *Chronicle of Higher Education,* "You're not giving a lecture on Foucault to a bunch of French theorists. You have five minutes to hit it or quit it."[1] The Associated Press says broadcast "actualities"—in your case, the time you're on camera or on the radio—should be long enough to enable the audience to establish in its collective mind who is speaking, but not so long that it loses its focus. How long is that? The Associated Press says ten to thirty seconds. According to Frank Kauffman, my former colleague now at Edelman Public Relations, sound bites for candidates in the 1968 presidential race ran, on average, forty-two seconds long—"an eternity on television today," he said. Today the average sound bite is ten seconds, plus or minus, and entire stories are told in that time.[2]

Ten seconds is not as short as it sounds. If you don't believe me, get a stopwatch and see how much you can say in ten seconds—when you are prepared. But it is not a lot of time. Don't waste it. Preparation is also important because so much television and even radio news is done on the fly. At an Aldo Leopold Leadership Program training session, Burnside offered a typical timetable for his broadcast day.

In the morning, he and his news director decide he will do a piece on a scientist who has just reported a new way to observe and measure rip currents. He arranges to interview the scientist at his research site, and drives there with his cameraman, in a truck equipped with satellite uplink and other equipment he needs. He arrives by midday and films his interview. There is not a lot of time to do things over.

Now that Burnside has the researcher on film, he edits his video in the truck—and with luck, by 4:45 he has the story down. He's on the air with it twenty-nine minutes later. Then he airs it again, with fixes made as needed, on the next hourly broadcast.

Meanwhile, news directors and editors have decided what the broadcast as a whole will contain. Typically, broadcasts are built from the top down—that is, the decision on what to lead with determines much of what the rest of the broadcast will be like.

The old television news adage "No picture, no story" isn't always true, but it is certainly true that television journalists value good footage and will actually cast their story according to what they can get on film. They seek synergy between words and pictures, Burnside says, "and you can only get that if you write to your video. If you are a good television reporter, you write to your video."

That's why you should think about what kind of video—animations as well as conventional film—you can provide. That's also why television journalists (and other journalists as well) will want to film you in your lab or in the field. Your institution's public information people can help you minimize disruption and can let you know if there are places or situations your institution will not allow to be filmed. (With rare exceptions, for example, the *New York Times* does not allow filming in the newsroom.)

Both radio and television are passive media. Perhaps radio engages the imagination more than television does, but many people will also be engaged in some other activity as they listen to or watch the news. And although they can go online to download and review many radio and even television reports, in general people need to get your message the first time they see or hear it or they will tune it out. To reach them quickly, and on the fly, you need a strategy.

Steve Curwood, the producer of the public radio program *Living on Earth,* says he develops his interview strategy by thinking first about his audience.[3] His ideal listener, he says, is a curious junior high school student eager to learn about science. Others in television say their mental audience is collectively about as savvy as the average tenth-grader.

To get through to this listener/viewer you must keep in mind lessons you have already learned: The sound bite is your friend. Figure out your one or two crucial points and make sure you express them clearly. Remember that the average person can speak (and easily understand) about 100 words per minute. Use short sentences and avoid technical terms unless you absolutely have to use them, and in that case define them as simply as you can. If you have to talk numbers, use simple, easy-to-understand figures. Again, percentages are easier to understand on the fly—but avoid the relative/absolute risk problem (remember, a risk that doubles from one in ten million to two in ten million is not the same as a risk that doubles from three in ten to six in ten).

If questions come quick and fast, answer one at a

time. If a good response does not come readily to mind, buy time by rephrasing the question—possibly in a way you would prefer to answer it. Do *not* try to avoid the issue; that will be obvious.

Dispel misconceptions immediately. Then offer the correct information.

YOUR VOICE

Do you speak distinctly? Tape yourself speaking and listen to the tape. Do you slur your words? Mumble? Strive to break yourself of these habits. Again, enlist a good friend to give you candid feedback. Listen to your voice. Is it compelling? Practice with your tape recorder.

Before an interview, warm up your voice by talking, even singing. Speak in a conversational tone and vary your tone and timing the way you would if you were telling a joke. But be deliberate—perhaps slow your pace slightly—when you are making key points. Don't let your sentences trail off and don't drone on and on.

If you are being interviewed on the telephone, stand up and walk around. You will sound more animated on the air. Even if no one can see you, smile. Believe it or not, it comes across.

The interviewer or producer will undoubtedly make this point, but I'll make it here too: avoid using cordless or cellular phones. Their connections are not as good.

See if your institution has a studio for television or radio broadcasts and arrange to use it if necessary. Or you may be asked to travel to a radio studio in your city for an interview. If you can, agree to this request.

Don't allow distractions. If you are being interviewed in your office, close the office door. Focus on the interview, regardless of what is going on around you or how much you want to read your e-mail. Use the bathroom before the interview, and have a glass of water on hand.

Most important, don't let your nerves keep you from the stage. Conquer (or at least reduce) stage fright not just by boning up on your subject but also by reviewing your sound bites and otherwise making yourself ready to be effective. Nervousness can actually be helpful, in that it can imbue your appearance with extra energy. And it diminishes with practice. Once you have appeared on television, say, and been happy with the results, you will be less worried the next time. Remember, not everyone has charisma, but everyone can learn to present information effectively.

YOUR DEMEANOR ON FILM

Be aware that you cannot necessarily tell whether a camera is filming, where it is focused, and, if it is pointed directly at you, whether the frame includes your face or your whole body. So maintain an "interview attitude" from the moment you are lit until the lights go off. If a camera is near, assume it is filming.

Don't slouch. If you are sitting, sit up straight and lean forward slightly in your chair. You will look taller and thinner than if you sit back. Try not to fidget, twitch, swing your legs, yawn, sigh, close your eyes, or roll your eyes. Use your hands the way you would if you were speaking to someone across your desk, or at the dinner table. If you gesture too broadly, your hands may wander outside the image frame.

Look at the host/interviewer, even if he is looking at the camera. Don't look at the camera. Refer to the interviewer by name.

If you are one of several guests on a television program, try not to sit between the interviewer and another guest or between two guests. If multiple people are being interviewed and you feel you have something to add, take advantage of a natural pause to say, "I'd like to comment on that as well," or "I'd like to add something here."

Don't worry if there are moments of silence. It is the interviewer's responsibility to keep things moving. Don't feel obliged to fill any empty spaces. You may find yourself saying something you later regret.

Never lose your temper. Don't take things personally, no matter how personal they are. If you are involved in a messy situation—accusations of malfeasance or the like—practice answering unwelcome questions. A good public relations person can help you. One possible approach: start off with, "Let's put that into perspective."

Try not to lose patience with reporters, even if you think they are spectacularly uninformed. Anticipate simple (stupid) questions, and figure out how to use them to convey useful information.

If you are seated for your interview, stay seated until the interviewer says you are off the air. And when the interviewer thanks you for being on the program, offer a simple "thank *you*" in reply.

YOUR LOOK

You don't have to have a television wardrobe, but some types of clothing definitely work better than others. You should avoid wearing busy patterns and solid black or white (black absorbs too much light; white reflects too much light). Don't wear stripes; they appear to shimmer on television. Ditch any flashy jewelry, even lapel pins. Don't have bulky stuff in your pockets.

Unless you are to be filmed in the field, a dark suit and blue shirt are best for men. Knee-length socks are always best, to avoid exposing bare leg when seated. To make sure your jacket will not ride up, sit on the tail of the jacket.

Wrap dresses work well for women—and anything V-necked is good for attaching a microphone. If necessary, sew the V-neck closed at the bottom or wear a camisole. Make sure your nails are neat and your hair is well combed and brushed, or even blow-dried for the occasion. As for wearing makeup, if the television people offer it, you are probably wise to accept.

If you wear glasses, keep them on. Fiddling with them, even putting them on or removing them, is distracting. Try to keep your coughs and sniffs off the air.

Avoid wearing anything revealing or outré. Nancy Baron of the Communication Partnership for Science and the Seas often shows scientists in her seminars a film in which a researcher acknowledges she was wrong to turn up for an interview wearing a T-shirt with the message "When all else fails, manipulate the data." That is an extreme case, but even the name of your favorite team on a sweatshirt can be distracting in an interview. If you are wearing a T-shirt, make sure it's a T-shirt free of messages.

If you are wearing an earpiece, be sure it is working before the interview begins. (If it should suddenly go dead in a live interview, signal the interviewer by tapping it.)

TELL A STORY

In the end, telling science stories on television and radio is just that—telling stories. I was struck by this (obvious) point when Paula Apsell, executive producer of the public television series *Nova,* came to speak to my seminar at Harvard. When she considers topics for her program, she said, she looks for "stories, not collections of facts." An ideal subject, she says, combines "a good story, strong characters, strong visuals, and a different point of view." It needs "some sort of conflict, mystery, obstacles to overcome." And it needs to be presentable on screen—it must have "a visual dimension in front of you, as opposed to in your mind, as with the printed page."

She was echoing something I had heard often before, in the *Times* science department and from another seminar guest, Gareth Cook, a Pulitzer Prize–winning science reporter for the *Boston Globe.* Cook said he and his colleagues at the *Globe* strive to avoid "term papers." At the *Times,* we say we work to avoid writing "encyclopedia entries." We are all looking for stories.

Apsell told the class that *Nova* documentaries, like

television dramas or movies, typically have three acts—what Apsell called the setup, the conflict, and the resolution. People who approach their subjects with this framework in mind, she said, can end up "making a story out of a collection of facts."

9

Telling Science Stories Online

The first time I heard a serious researcher seriously suggest that scientists should post material on YouTube, I thought she was joking. (I was so unable to take the idea seriously that I don't even remember who made the suggestion.)

Then, in the summer of 2008, the European Center for Nuclear Research, CERN, opened a particle accelerator called the Large Hadron Collider (LHC), the biggest apparatus yet built in the quest to unravel the secrets of matter and energy. Hours of television time and tons of newsprint were expended in the effort to tell people what would go on at the multi-billion-dollar installation, and what this work signified for our understanding of the universe.

But perhaps the most intelligible and entertaining explanation of the enterprise came not from an august science journalist but rather from a twenty-three-year-old CERN science writer, Kate McAlpine, who raps under the name Alpinekat. Her LHC rap, viewable on YouTube, offers a rhyming description of the search for a hypothesized subatomic particle called Higgs boson and an explanation of why anyone wants to find it. (In theory, it endows matter with mass.)[1]

McAlpine recruited fellow CERN workers to sing and dance on camera, and created a rap that explains the workings of the machine. Physicists I know who have seen it describe it as accurate—certainly "accurate enough." A CERN spokesman has described the rap as scientifically "spot-on."[2] And it is for sure far more entertaining than the staid "official" explanations.

In a commentary she wrote for *Symmetry* magazine, McAlpine calls rap "a good way to communicate. Rhyme has always helped embed words in my mind; hopefully science rap can help cement ideas in the minds of students and other interested people."[3]

McAlpine's LHC rap offers a vivid example of how scientists can use the Web as an instrument of enlightenment.

Just as television started out as radio with pictures, the Web started out as words with embellishments. Today, though, telling stories online is not simply a matter of telling them with words, or with words, hypertext, links, photos, and so on. Though the vernacular of the Internet includes the vernacular of print and of broadcast, its own vernacular form is quite different. You will need to learn how to communicate in this form if you want to succeed in getting messages across online—either as a source for someone else's site or for your own Web site or blog.

This new vernacular is demonstrated to me every time I ask students to read a daily newspaper—the *New York Times,* of course. Most of them say they are already daily readers, online. But then I ask them to read a copy of the *paper* paper, and to note what headlines, articles, graphics, and captions they read, how far into them they read, and so on. After this experiment, most of them say the same thing: they read more stories on paper, but in less depth. Or to put it another way, they read fewer items online, but they read them in more depth—opening links from them and so on. Uniformly, they are astonished at the variety of subjects in the printed paper (including advertisements) that they never noticed online.

I don't know what this says about the future of journalism. In fact, I am worried about it. I like the serendipity of opening the paper and discovering myself reading something I would never have bothered clicking on, on a Web site. I worry that sites that cater to readers' known tastes will discourage the kind of accidental discovery that has been a hallmark of newspaper reading.

Also, online journalism is developing standards that differ, sometimes wildly, from what some journalism scholars call "the discipline of verification," a hallmark of the mainstream media. That discipline can give way online to a "journalism of assertion" in which people post, often anonymously, erroneous, defamatory, vulgar, or mindless observations that would rarely, if ever, gain attention at a respectable news outlet.

But let's leave those worries aside. Though there are a few holdouts, the movement of news and information onto the Internet is inevitable and accelerating. Consider how you can use the Net effectively to communicate not just with fellow researchers but with the wider world. The LHC rap shows that the Web offers great scope for your imagination, but it also offers great opportunities even for those of us whose frame of reference is more conventional.

As a writer and editor who has spent a career trying to cram way too much information into too few newspaper column-inches, I welcome the Web's capacity to accommodate extra copy. Plus, my words can be accompanied by photographs, slide shows, recorded interviews, Flash-based graphics, primers, and blogs on which readers can voice their own views, respond to polling questions, and participate (with me) in online forums.

But though the medium's capacity is in theory infinite, the audience's capacity is not. So stories posted online must still adhere to the same kind of space discipline that marks journalism on paper.

Also, although the *New York Times* posts articles online more or less as they are written for the newspaper, many people believe writing explicitly for the Web calls for different techniques. On paper, about three-quarters of readers who start an article on page one leave it at the runover, where the article jumps to an inside page. So far, it looks like writers probably lose just as many readers online each time an article is interrupted by a subheading and particularly by an ad—and advertisers like to put their messages in the middle of articles.

Also, the rambling, narrative introduction—the so-called anecdotal lede beloved by many journalists—does not seem to work well online. It is better to write in short sections that can be set off with their own headings.

For example, when Ken Weiss began his work with Usha McFarling on their prize-winning five-part series on the oceans for the *Los Angeles Times,* he knew from the outset it would appear in the paper and also online. He learned, he said, that readers have, on average, an attention span of about two minutes for any one item online. "So we broke it into two-minute segments," he said.[4]

I received similar comments and advice from Web-content producers at the *New York Times.* For example, they say that unless you have truly compelling images, eight to twelve is probably the maximum for an effective slide show. If you accompany the images with a voice-over or other audio, you can raise the number to ten to fifteen. In our online slide shows, each image is on screen for eight to ten seconds, more or less, depending on how easy it is for viewers to understand or "read" the image.

I can already hear researchers objecting that it is pointless to try to communicate seriously with people whose attention span limits them to a minute or two per item. But if you are interested in reaching an audience, you must consider the capacities of that audience.

Other online guidance comes from the EyeTrack project, a series of studies conducted by researchers at various institutions in conjunction with the Poynter Institute, a journalism education and research organization. The researchers use special glasses and other equipment to track what people focus on when they scan online news sites. So far they have found that people read more when the type is presented in one column down a page, rather than two or more columns across it. Just as on paper, shorter paragraphs do better—forty-five to fifty words, max. Large type encourages readers to scan; they are more likely to stop and read smaller type. And, no surprise, readers seem to absorb new information better when it is accompanied by graphics—charts, drawings, and the like. The bigger the image, the more time readers spend looking at it.[5]

Some of these reading patterns may be hangovers from the days of paper newspapers. It is probably too soon to tell. In fact, it may be a while before we know what works best for telling stories online.

WEB SITES

If you don't have your own Web site for posting news about your research, find out if your institution will help you set one up on its site. If you want to strike out on your own, and if you don't have the time or skill it takes to set up a Web site, there are companies and consultants you can hire to design a site for you—for fees starting at perhaps a few hundred dollars. For an additional cost, they will maintain and update your site for you, with material you provide. Google "Web hosting" for information about companies that offer these services. You can also create your own site using design software that you can download.[6]

Jeremy B. C. Jackson, a marine biologist at the Scripps Institution of Oceanography at the University of California, San Diego, led an online effort, called Shifting Baselines, to draw attention to the slow, steady diminution of the natural world. Another group of researchers, including climate expert and White House science adviser John Holdren and Jane Lubchenco, the marine ecologist tapped to head the National Oceanic and Atmospheric Administration, have launched an online site called Climate Central, designed to be an informative and accurate source of climate information for the news media, public officials, religious leaders, and others.[7]

Keep in mind that if you have a Web site for your scholarly work, you may be interacting with the public there whether you want to or not. If your site turns up on search engines, nonscientists will see it. When you think about what you will post on your site, think about not just the researchers who will read it but also the people outside your field who may turn to it for information and guidance. You can be clear and engaging without sacrificing intellectual rigor.

You may be tempted to post videos on your site. Take

a critical look at them before you do. They don't have to be Hollywood smooth, but they should not be out of focus, under- or overlit, photographed against cluttered or confusing backgrounds, or otherwise just plain sloppy. You don't want small technical problems to overshadow your message.

But before you get that far, you should consider why you are embarking on this venture. Are you setting up a Web site to provide information for fellow researchers? To help members of the public interested but not expert in your field? To aid students with their projects? Or are you looking for a soapbox for your own opinions? How you answer that question will help you shape your site.

Consider your audience. How tightly defined is it, in terms of education, experience, or expertise? Will you need to target different parts of the site to different knowledge levels?

Think hard at the outset about what the site's tone should be, and how that tone can be established by typography, color, and other elements. Design secondary pages with a shared common layout and typography. If this kind of design is not your strong suit, get help. And if the help comes from a colleague and not a professional Web designer, be sure that the colleague has a Web site you admire and find useful.

Certain typographical and writing practices increase readability online. Among them:

Bulleted lists of major points and examples

Occasional use of bold type

Subheads

Writing about one idea per paragraph

Using less verbiage overall

There is growing agreement that people who read online are not reading in the dead-tree sense of the word. That is, they are not reading for fun. Rather, they are foraging—searching for key words that relate to whatever they are researching. To reach them, be obvious.

Your site must be easy to navigate, especially if it is large or complex. People seem to have limited patience for Web sites—if they don't find what they want easily, they will turn to another site. And research so far suggests that people prefer to find navigation guidance at the top of the page. And of course your site must be searchable.

Consider your title. Titles can make or break your site in terms of Web search engines.

Offer useful links—to sidebars, video and audio presentations, even other Web sites. Be aware, though, that deciding where to link is more complicated than it seems. In the science department at the *Times,* when our Web presence was relatively new and small, we had several discussions about whether linking to other Web sites from our pages on the newspaper's site suggested to readers that we endorsed the sites. I feared that it did, despite on-screen warnings that they were leaving the newspaper's sacred electronic precincts. As a group, we never did agree. When I asked my colleagues what sites we should link to, opinions ranged from "all of them" to "none of them."

Today I think people are used to moving from one site to another and understand that the standards of one are not necessarily the standards of another. Also, visitors to your site may want to know what's being said on other sites, even if they trust your page and not the others.

Make sure your site tells people how to reach you—perhaps at an e-mail address you set up expressly for this purpose. A snail-mail address is good too, and, if you have the nerve, a telephone number.

BLOGGING

Sheril Kirshenbaum, a researcher at the Nicholas Institute for Environmental Policy Solutions at Duke University, says she became a blogger by accident, when she told students who had been urging her to blog that she would start when the Democrats regained control of both the U.S. House and Senate. In 2006, they did.

Kirshenbaum started by posting on other people's blogs "one or two times a month," she said.[8] Then her friend Randy Olson, a biologist turned documentary filmmaker, introduced her to Chris Mooney, the proprietor of ScienceBlog. He asked her to take over for a week, she recalled, because he was taking his first vacation "after six years of blogging."

"That week led to another trajectory," Kirshenbaum said. "He said 'stay,' and here I am, I am a blogger." She now contributes to ScienceBlog and the Wired Science/PBS blog Correlations.

Judging from the number of people who have clicked on her postings, she said, "The impact is immense." As a blogger, she said, she can "communicate with people about the questions they have in science. It is a new way to communicate, immediately."

Advocates of blogging say the blogosphere is the ultimate marketplace of ideas in which good thinking drives out bad. As Kirshenbaum put it, "If someone starts writing things that others don't buy, they are going to get slammed on it."

People like Kirshenbaum say it remains to be seen when it will be "the norm" for scientists to blog. It also remains to be seen how much time anyone will have to read the blogs, or anything else, if everyone is blogging. ScienceBlog sweeps the science blogosphere and offers links to useful or interesting sites. Other sites use algorithms of one kind or another to aggregate news by topic area and so on.

Kirshenbaum points out that, for scientists, "by writing and engaging and participating every day, we show the public we are able to talk about bigger questions."

According to Scott Gant, whose book *We Are All Journalists Now* is a fascinating discussion of how the advent of the blogosphere is changing ideas about journalism, there were 9 million bloggers in the United States in 2005, and another 40,000 blogs were being started every day. By 2008 there were by some estimates 100 million blogs in the world, of which about 15 million were active.[9]

Some say scientists hang back from blogging, "possibly because of the association of blogging with Internet chat-rooms or even the fear of intellectual property theft," as Alison Ashlin and Richard J. Ladle of the Oxford University Centre for the Environment put it in a Policy Forum in the journal *Science*.[10] Are they right? I think it's too soon to tell—in part because blogging is both easy and difficult.

Establishing a blog can be as simple as visiting one of the online do-it-yourself blog sites, like Blogger. You can even sign up with ad services that will place ads on your blog and pay you a fee whenever a reader clicks on one of them—assuming you meet the service's minimum monthly traffic quota. (Don't expect much.) But producing a blog means performing most of the tasks required to maintain a Web site—providing material, updating it, adding useful links (and deciding which are appropriate), and so on. So if you want to take it on, be sure you are in love with your subject.

Other good advice comes from, of all people, the columnist E. Jean Carroll who writes for the magazine *Elle*. She says the number one career topic her writers ask about is blogging. A while ago she replied to a reader—"with a blog (just like everybody)"—who wanted to know how to draw more attention to it.

E. Jean's advice is probably good, but it is daunting, even if you ignore her first two suggestions: that you insult "a blogosphere icon" and hope that "lathered-up commentators will produce links to your blogs," or that you launch a video blog, or vlog—and post while naked.

Her more serious points: Own your subject—that is, be absolutely up to date on everything having to do with it. And do the work. As E. Jean put it, "Break stories; update with new links, pictures and videos; and for gawdsakes write with style."[11]

On your own blog, you can post whenever you want to—hourly, daily, weekly, whenever. But some experts say maintaining a regular rhythm helps draw readers.[12] Others advise keeping your posts to the same general length. (It is not intuitively obvious to me why this should be helpful—again, time will tell.)

The etiquette of the blogosphere demands that you link to other bloggers whose work you cite. The advantage of this is that your links to them may encourage them to link to you. You can try plugging your own blog in comments you post on other people's blogs, though some don't allow it. Kirshenbaum links her blog entries to DotEarth, the blog Andrew C. Revkin maintains at NYTimes.com, and vice versa.

Perhaps the knottiest question is whether to allow people to post comments on your blog. If you do, you can in theory create a kind of community-building device, a place for people to exchange views on important or interesting questions related to your work. Plenty of researchers say they have found this kind of blogging environment immensely helpful.

But it is now a truism online that as a blog draws wider attention and the threads of reader-posted comments grow longer, the probability of their degenerating into vicious nonsense approaches one.

If you allow people to post on your blog and you don't monitor their postings, you will have to accept the possibility that it will end up with readers who participate "in coarse, bullying and misinformed ways," as Clark Hoyt, the public editor (or ombudsman) of the *New York Times,* put it in a column on the newspaper's blogs.[13] Hoyt defended the newspaper's decision to monitor postings and remove those that are intolerant, irrational, or the like. If you take steps to remove this kind of thing, though, you will certainly be accused of imposing unwarranted restrictions. "Tepid civility" is the term one poster used to describe the tone of a *Times* blog.

You may also be blessed with people who post to your blog so often they risk taking it over. In this case, if you limit their posts you may be accused of censorship.

Even correcting errors on the blogs of others can be time-consuming. But despite the time they require, as Ashlin and Ladle put it in *Science,* "If environmental scientists ignore online communication platforms such as weblogs, we run the risk of creating a generation of eco-illiterate consumers and voters at a cruel time for Earth's diminishing resources."

A few people actually earn a living blogging. But that

depends on driving a steady stream of traffic to your site. That, in turn, depends on a steady flow of new material. Maintaining such a pace is difficult. After running his blog DotEarth for a while, Revkin began describing living in the blogging world as being in a room with a giant balloon that expands and expands until it seems to occupy all the space of life.

Even people who led what ordinary people would describe as busy, complicated lives before turning to blogging describe its nonstop nature as unrelenting.

Perhaps because of this pressure, a new approach to blogging is modeled in a way on the "slow food" movement. It's called "slow blogging."[14] One proponent of slow blogging is Todd Sieling, whose "Slow Blog Manifesto" (motto: It happens when it happens) is posted online. As he sees it, slow blogging rejects immediacy in favor of "speaking like it matters" and reflects "a willingness to remain silent amid the daily outrages and ecstasies." Slow bloggers do not care how many people read their pages. Possibly as a result, most slow blogging sites tend to attract few readers. Mid-rank sites, "among the more visible bloggers," may attract a few hundred readers.[15]

So with blogs perhaps the biggest question is, how much time and effort do you have for them, either your own or others? There are already way more blogs and blogging threads than the world needs. So perhaps E. Jean's best advice is think twice about becoming a blogger: "If you don't possess the stamina, write Wikipedia entries instead, my luv, and enlighten people."

10

Writing about Science and Technology

Daniel Pauly, former director of the Fisheries Centre of the University of British Columbia, used to fume about the poor quality of writing on scientific and technical subjects. "I could do better than that," he would think. But then he accompanied a colleague to a karaoke bar, and his outlook changed.

What happened was, his colleague decided to sing. The would-be vocalist had a microphone, musical accompaniment on the karaoke box, the lyrics to his chosen song, a dark and romantic atmosphere—even an intoxicated audience. Still, he bombed.

Somehow, Pauly told me years later, this episode

made him think about how information is communicated. It is not enough, he realized, to know the language and understand the subject. Something more is needed to put a story across—the skills of the journalist.

It would be silly to suggest that writing about science or engineering is comparable to singing in a karaoke bar, or that reading a chapter in a small book like this one could turn anyone into an accomplished writer. I believe there is no way to become a writer other by than reading other people's work and practicing your own.

But I can offer some useful suggestions for people interested in writing for a lay audience. After all, though you may need inspiration and talent to be a poet, it is not difficult to become an adequate writer whose prose is clear, concise, and engaging.

Unfortunately, though, scientists and engineers tend to start with some handicaps. Often they have managed to escape formal training in writing. Their subject matter can be complex, arcane, and difficult to convey. They may be skilled at analysis but uneasy at synthesis—the kind of summing up and conclusion drawing that good writing thrives on. And, of course, some researchers take perverse pride in a kind of hyperaccuracy that can fill their prose with so many caveats it becomes practically incomprehensible.

"It is hard for authors to avoid aiming reports of

original research at the cognoscenti, especially in fields where movement at the frontier is active," wrote Donald Kennedy, then the editor of *Science,* in an editorial in that journal.[1] "Because the methodological grain of each discipline has become extremely fine, it requires heavy use of technical language, jargon and acronyms. That tends to make even the title of the average communication in molecular biology in any top-tier journal impenetrable by an ecologist, let alone a physicist."

David Damrosch, a literature professor at Columbia, calls this phenomenon "coterie writing"—writing for a small circle of like-minded readers.[2] Coterie writing is heavy on jargon, acronyms, abbreviations, and other verbiage specific not just to your field but to your own small circle within it—your coterie.

"It is neither necessary nor desirable to dumb our projects down when writing for a general audience," Damrosch wrote in the *Chronicle of Higher Education.* But, he said, "we need to write quite differently when we want to reach beyond the comforting confines of our disciplinary coteries"—to people who don't understand the subject and may need to be convinced that it is worth reading about.

My *Times* colleague James Glanz, a PhD physicist as well as a journalist, once told me of a notice he saw in *Physical Review Letters,* a top-tier physics journal. He said it instructed would-be authors that the first three paragraphs of every paper had to be comprehensible to any garden-variety PhD physicist. I tell this story to scientists and engineers to illustrate the difficulties we journalists face when trying to figure out what's what in the world of research. Setting the comprehensibility bar that low does not encourage researchers to develop the habit of writing clearly.

Perhaps that is why Vernon Booth dedicated his book *Communicating Science* to a mythical reader he named T. W. Fline—for "Those Whose First Language Is Not English."[3] He urges scientists to write so clearly that T. W. Fline will understand their work. It is a good plan to keep a reader like this in mind. It does not have to be Ms. Fline herself—it can be a neighbor, your grandfather, an inquisitive twelve-year-old, or anyone else you think ought to be able to understand your words.

Bring the same respect to your lay audience that you would bring to an audience of scientific peers. Think about your material from the perspective of those you hope will understand (and enjoy) it.

"Good science writing has the audience firmly in mind," Roald Hoffmann wrote in *American Scientist.*[4] Hoffmann, a founder of the Cornelia Street Café in Greenwich Village, a "café scientifique" where members of the public can meet and discuss developments in science, says writing about science for a lay audience can help scientists write better for their peers. Writing for a general reader "sets free the oft-suppressed metaphor," he says.

A good way to improve your writing is to read what other people are writing, in print and online, and to watch and listen to what is on the air. Notice what you think works well, and try to figure out why. Know whose work you like and think about why you like it. (Actually, if you are not already doing this kind of thing, reconsider your desire to be a writer—it's just too hard for people who are not really into it.)

Good writing is clear. Assess your own by stepping back and looking at every specialist word and concept in the piece. Ask yourself who knows what it means and who does not. Use straightforward language and verbs in the active voice. (If you don't know what I mean by this term, buy a good usage manual. You need one.) Use relevant, engaging examples and analogies, and offer concrete examples.

There are plenty of guides for would-be writers (some of them are listed in Suggested Reading in the back of this book). Another place to look for writing guidance is the Web site of the *Times,* which regularly posts After Deadline, a column that contains discussions of our in-house guidelines and advice for "smoothing the rough spots" in prose.[5]

My *Times* colleague Natalie Angier, an accomplished writer, says writing about science is what writing about baseball would be like if you always had to define terms like *home plate* or *base on balls*. So it is. When you start to write a science story, imagine how you would describe your favorite baseball game if your listener had never seen baseball played.

That does not mean dumbing down your writing. As actor Alan Alda says about his experience on the public television program *Scientific American Frontiers,* which he hosted for years, audiences are quick to discern when you are patronizing them—and they don't like it. They will tune it out. Clarity is not the same as dumbness.[6]

State your conclusions in a way that is interesting and informative, but not dogmatic. Remember, you are not writing a textbook to be read by students who will be tested on what they absorb. You are writing a story, something you hope will engage the imagination of your readers.

As with your sound bites, avoid jargon. How do you know if a word is jargon? Again, think about whether it's a "headline word." If it's not, consider whether you can tell your story in lay language and leave it at that. If the term is crucial, define it first. And just as you would with sound bites, remember your metaphors, similes, and analogies.

Most important point: tell what you need to tell to move the story forward. No more, no less. As Angier put it in an article in *Science Writer,* "You don't have a huge and devoted following" hanging on your every syllable.[7] When in doubt, leave it out. When I am writing, I try to think of my story as a cart moving along a track. Every word is either propelling the cart forward or weighing it down. I try to ditch the dead weight.

This kind of cutting can be hard to do. It is not always intuitively obvious which details are crucial and which are clutter, especially if they are *your* details and you find them fascinating. Resist the temptation to shoehorn unnecessary information into your piece just because you think it's cool.

In his book *Writing with Power,* Peter Elbow describes his writing method: give half your time to writing and half to revising. He recommends just sitting down and writing everything that occurs to you that pertains to your theme. Don't repeat, don't digress, and don't worry about order or wording.[8]

When it is time to revise, bring your audience to mind and then read through what you have written.

Find the best part or main point. Rearrange other points around it.

When I am writing an article, unless it is a very simple, short spot news story, I usually make an outline. A typical outline might look like this:

1. The lede (top) of the story. Sometimes that's a bare recitation of the most important fact or finding; sometimes I start off by telling a story, as I have done with some of my chapters here (the anecdotal lede).
2. The answers to the all-important questions: Why are you telling me this? Why now? Journalists call this essential paragraph the "nut graph" or "cosmic graph." It might say something like: "The finding is important because" or "for the first time researchers can explain . . ." It does not have to come immediately after the lede, but readers should not have to wait long for it.
3. Material to explain and amplify the lede.
4. Necessary background. (Necessary!)
5. Supporting material.

If you are writing on a computer (and who is not?) you can write your story into your outline. Add missing portions, delete extraneous portions, figure out your conclusion (your "kicker"), and produce a draft. Read it through again, looking ruthlessly for anything that can be removed.

Consider your audience. If you are writing for the kinds of people who read Science Times, say, remember that surveys show that these readers—ordinary Americans—will find sentences of seven to eleven words easy to read and sentences of more than twenty-five words hard to read. At the same time, don't turn your prose into a stream of short, choppy sentences. Varying their length makes for more rhythmic prose. The same is true of paragraphs—they should not all be the same length.

Keep subjects close to their verbs. Put relative or dependent clauses where they won't interrupt the subject-verb combo. If you don't know what I am talking about, consult a good usage manual. (Again, you will find some suggested titles in the back of this book.)

Avoid the verb "to be." Use action verbs whenever you can. You don't have to follow this rule out the window, but it is a good rule. And use the active voice. Don't say, as Paul Revere might have, "An approach by the British has been observed." Instead say, "The British are coming!" (Actually, he called them "the regulars," but this is the kind of showing off you should also resist.)

For some reason, scientists notoriously use the passive voice—things are added, are measured, are found, and so on. Using the active voice forces you to be specific about who added, who measured, who found—and can point to holes in your story.

This next suggestion is perhaps my strangest, but it works: when writing in English, a language derived from German, strive to use words with German roots in preference to words with Latinate roots. Talk about cats, not felines, or water that is safe to drink, rather than potable. Don't inhale and respond, take a breath and answer.

Embrace the use of narrative—when you can. If the work you are writing about naturally has a narrative spine—years of research pay off in a notable discovery, say—don't be afraid to tell it as a story. As Hoffmann put it, "Human beings prefer to organize their hard-won knowledge of reality in the form of a story."[9] So think of your piece as a story with a beginning, middle, and end, drawn together by the power of its narrative spine.

One important caveat: it is possible to fall so in love with a narrative that you will be tempted to omit, include, or twist facts for the sake of storytelling rather than information conveying. Needless to say, you must avoid these impulses; in my opinion, they are a particular hazard in science writing. Another caveat—and one that ought to be unnecessary for a researcher: remember that the plural of anecdote is *not* evidence.

Avoid euphemism. Especially avoid a euphemism someone else has adopted to make a political point, terms such as "pro-life" or "healthy forest initiative." Call things what they are—"opposition to abortion rights," or "a proposal to increase loggers' access to federal lands."

Avoid clichés. If you watch like a hawk (ahem), you can spot them.

Avoid puns and wordplay. This advice is a matter of taste, of course, but in my experience puns and other wordplay are usually the refuge of writers who do not want to spend time figuring out how to say effectively what they want to say. Here is what I tell my writing students: assume you have a lifetime allotment of one pun. If the pun you want to use is worthy, go for it. If not, say it another way.

Think about whether your piece would be better if it were accompanied by a photograph, drawing, chart, map, or sketch. If you cannot produce these graphics yourself, be ready to offer suggestions to your editor.

Finally, reread and rewrite. Then, stop rereading and rewriting. Prose is like pastry—after it has been handled too much it can start to become stiff. Practice will teach you to know when you have improved your prose "enough."

There are a few things you can do when writing to

save yourself from trouble after your piece is published. For example, whether you are writing a news article, a feature, or an opinion piece, inoculate yourself against criticism by acknowledging empty spaces in your argument, gaps in your data, or findings that are less than show-stopping. If others disagree with you, admit it. Then say why your position is stronger.

Be fair. If you are going to hammer someone in print or online, give that person a chance to reply. If you think you may be venturing into actionable territory—if you may be defaming someone—be careful. The *Associated Press Stylebook* contains a good, quick review of libel law. When in doubt, obtain advice from a public information officer or someone else who actually deals with the media.

Unless you are discussing something of your own knowledge, say where your information comes from. Remember, your most important attribute as a writer lies in the willingness of complete strangers to believe it's worth paying attention to what you have to say. Protect this credibility by not putting your name on any information you are not personally sure is correct. The Associated Press used to tell writers to attribute anything more in doubt than the fact that the sun rises in the east. You don't have to go that far, but you should be careful.

Finally, and this is a skill scientists seem to have real

trouble picking up, be willing to be edited. Accept that you might not be the best judge of your own prose, what works best for the publication you are writing for, and so on. It is very common for writers to resent editors' questions or suggestions, but resentment is unhelpful. If someone reads your piece and is left with questions, step back and ask that person what she does not understand. You don't necessarily have to accept every change your editor proposes, but before you start to argue, find out what motivated the question or suggestion.

What does editing entail? Some of it is straightforward: correcting grammatical and typographical errors, for example. Though you should consider yourself responsible for the accuracy of your prose, your editors may check for errors of fact as well. They will also make your prose conform to the publication's "style"—its rules about which words are abbreviated and which are spelled out, for example. And if necessary, they will trim your work to fit its allotted space. (If you think they are cutting the wrong things, explain why and offer to make your own cuts.)

If the editors think your prose is choppy, they may suggest transition phrases to smooth things out. They may tell you your essay is unclear in places, or ambiguous in ways you never thought about, or that it makes logical leaps unsupported by facts. They may even challenge some of your assertions—especially if they seem to be contradicted by facts or could be regarded as defamatory.

These kinds of comments and suggestions do not mean your editors want to weaken your arguments. Usually, it means they want to help you make your points more effectively. Don't be bullied, but don't reject their suggestions out of hand.

Instead, consider the experience of Orrin H. Pilkey, Jr., a coastal scientist who has written several books on coastal land use and erosion hazards. Of his experience working on his first book, he said, "I remember constantly squabbling with the editor, who kept saying it was too technical, too complex. I thought she was some kind of ninny who didn't appreciate good science."[10]

After the book came out, he was "really happy with it," he said, until he ran into some old friends, both highly educated but not scientists or engineers. "They complimented me on the book and then made the startling statement, 'Too bad it's so technical.' What a blow. After that I paid attention to the editors' criticisms."

A few months before his death, I attended a talk by the author John Updike, who was introduced by my *Times* colleague Charles McGrath, a former editor of the *New York Times Book Review* and former fiction editor of the *New Yorker,* where Updike was a regular contributor for years. McGrath told the audience that while Updike was an accomplished writer, he was also good at being edited—he understood that the writing process included the editing process and he was a willing participant. "Some writers get defensive," McGrath said. "They shut down."

Don't be like that. Follow Updike's example.

If an editor makes a suggestion and cannot defend it adequately, fight it. But pick your fights. If the suggestion does not change the sense or emphasis of the article, maybe you shouldn't fight too hard against it.

THE INTERVIEW

When you are not writing about your own work or your own ideas—and sometimes even when you are—you may need to interview others about their work, their findings, their opinions, and so on. Interviewing skills improve with practice, but there are some steps you can take ahead of time to prepare.

First, think about what information you are hoping to obtain from the person you will talk to. Write down your questions. When I am interviewing someone in person, and taking notes, I write my questions on a separate piece of paper I can refer to without flipping pages in my notebook. When I take notes on the computer, during a telephone interview, say, I write the questions into the file I will use for the notes.

When you call or e-mail someone to request an interview, tell the person briefly what the interview will be about, what you expect to produce with the information (an op-ed piece, for example), and how much time you think the interview will take. If you find yourself in the uncomfortable position of preparing to write something potentially negative about the person, be up-front about it. Tell the person what you are writing about and say you want to make sure he has a chance to make his views known.

Start with simple questions, like how to spell the person's name (always check this information, no matter how well you think you know it) and what his job title is. If you need to include this information, you must be accurate. And by starting with this kind of question, you also establish a pattern in which you are asking something and your interview subject is answering you. Some people are put off by note taking or recording during an interview, but answering questions like these can help get them used to it.

Next, strive to get your subject talking by asking open-ended questions, not questions that can be answered with a single word, phrase, or yes or no. For example, instead of asking, "Did you have a postdoc?"

try, “Tell me about your early years as a researcher.” Or instead of “How many graduate students do you have in your lab?” try “Tell me about your lab.”

StoryCorps, an independent, nonprofit project that records ordinary people interviewing friends, relatives, and others important in their lives and then puts the results on National Public Radio, offers some other good examples of open-ended questions that may help you think about how to frame questions for your interview:

> What was the happiest moment in your life?
>
> What are you most proud of?
>
> What are the most important lessons you have learned in life?
>
> How would you like to be remembered?

You might think that, if you are talking to a fellow scientist or an engineer, asking “What was the happiest moment of your professional life?” is unlikely to elicit useful information. But you never know. If you have the time, take the chance. Remember that people respond to stories. They absorb information presented in narratives. So if the person you are interviewing focuses on nuts and bolts, try something like “What motivates you to do this work?" My guess is the answer is unlikely to be anything as banal as "My paycheck."

If you ask technical questions and get a technical answer, put the answer in your own words, in lay language, and feed it back to your subject until he is satisfied with your phrasing. Don't end the interview until you can explain technical points simply and in a way your subject regards as accurate, or accurate enough.

Abandon your ego. If you don't understand something, say so. An interview is not an occasion for you to show your source how much you know, it is an opportunity for you to absorb information from your source. Don't be afraid he will leave the interview thinking you are a dunce. Let him be dazzled later by the brilliance of your article or broadcast. If you know a bit about your subject's field of study, don't assume he thinks the way you do about it. Ask.

If you are taking notes rather than recording your interview, don't be afraid to tell your subject he is talking too fast, to say, "Stop a minute so I can get this down," or to ask him to repeat something he has said.

In general, you should not alter ("fix") anything you present within quotation marks—ever. The *New York Times* allows us to correct inadvertent grammatical errors, false starts, and the like, but that's it. If a quote doesn't work as uttered, don't present it as a quote; instead paraphrase it, out of quotation marks.

BECOMING A PUBLISHED WRITER

Potentially, at least, there are many news outlets that might be happy to have a contribution from you: your local newspapers and magazines, alumni magazines or other publications at your institution, the newsletters and other publications of professional associations in your field, and so on—plus their Web sites.

But you should not expect editors to track your work and seek you out as a writer. If you want to write for a particular publication, you will have to let its editors know, a task you accomplish in what is called the "pitch letter."

If you are thinking of magazine writing, for example, you should figure out which magazines you might like to write for. The only way to do this is to read the magazines. Get an idea of what the magazine landscape looks like, and where you might like to position yourself.

Describe the story you want to write—and describe it as a story. Be engaging. Answer the "so what?" questions (Why are you telling me this? And why now?). Don't drown the editor in detail, but don't leave out strong points on the theory that she'll see them when you file the piece. She will never see the piece she never commissions. Tell her what kind of art could run with the piece; if possible send examples.

Note that magazines have much longer production schedules than newspapers and Web sites. If your story will be stale or outdated in a month, it's probably not good material for a magazine.

In your pitch letter, describe your qualifications for writing the story and send samples of your work. Keep a portfolio of your published work, Web pages, and so on, because this is the kind of thing editors want to look at when they consider commissioning writers. Remember Dr. Pauly's karaoke-singing colleague—mere knowledge of your subject is not, by itself, evidence that you can write engagingly about it.

Most editors prefer to receive electronic communications—e-mails are more quickly and easily dealt with. You can also follow up easily on your e-mail letter. If you don't receive a reply in a week to ten days, send a reminder.

If you really get hooked on science writing, you might consider getting some formal education in it—although, as someone coming from the learn-by-doing school of journalism, I don't think formal journalism education is crucial. But there are programs in science journalism at a number of universities, including the University of California, Santa Cruz, Boston University, New York University, and elsewhere. Such programs can provide valuable training, useful contacts, and the kind of writing credential some editors will pay attention to.

11

The Editorial and Op-Ed Pages

When journalists talk about the separation of church and state they are not ordinarily referring to faith-based initiatives or municipal Christmas trees. They are referring to the organization of the companies they work for—and the way the typical news department is separate from and insulated against the company's business operations and its editorial (opinion) writers.

Although many people find it hard to believe, the people who own reputable news organizations do not allow their own ideas or business considerations—chiefly, the wishes of advertisers—to influence the news coverage. For example, in 1999, when an executive from a cereal company, brought in to run the business side of the *Los Angeles Times,* made a special arrangement with an advertiser, the explosion of outrage in the newsroom forced him out of the company.

Reputable news organizations limit expressions of opinion to their editorial and op-ed pages or on-air editorial segments. These are also the venues where researchers can voice their opinions and engage their community.

LETTERS TO THE EDITOR

One point of entry is the letters column. Most print publications and online outlets have one, and even television and radio shows run letters from listeners, usually on a regular schedule. Letters are chronically among news organizations' most popular features.

Writing a letter to the editor can be a valuable exercise, even if your letter is never published or aired. The process trains you to express yourself tersely and clearly. Your letter can also help educate the people at the news organization you are addressing. The volume of letters received indicates to editors how much interest there is in a particular subject. And a cogent letter identifies you as a person with knowledge of and opinions about your subject, and the ability to express yourself clearly. It identifies you as a potential source.

But writing a letter to a news organization in hopes it will be published or broadcast is not the same as dashing off a note to your mom or an e-mail to a colleague. You have to know how to do it.

A first step is to become a regular reader/viewer of the letters the news outlet is already running. Get a sense of their length, their tone (which may vary a lot), and the kinds of subjects they cover. There is no reason why you cannot be the one to break new ground here, but be familiar with the landscape.

Then wait until something in the news inspires you to write. When it does, move fast. As with everything else in the news business, timeliness is crucial. Send your letter by e-mail, not snail mail. And write your letter into the text of the e-mail. Don't send it as an attachment. Because of security concerns, many news outlets instruct employees not to open attachments sent from strangers.

A successful letter is terse. Pay attention to the length of the letters the news organization typically runs and stay within range. Start by referencing the article or news event you are referring to. Someone who is coming to the issue through your letter should be able to tell what you are talking about. (If you are writing about a news item, include for the editors' information the date and page on which the article appeared or the date and time of the broadcast.) Then state your objection, clarification, amplification, related idea, or whatever it is you want to communicate.

Provide information on how to get in touch with you, your affiliation and title, other relevant information about you—that you are the author of a book on the subject, or the head of a professional association that deals with it, or the like. If your letter is to a radio or television outlet, say how to pronounce your name.

If you are a published author, note that letters can be an effective way to promote your books, but note also that editors are wary of letters whose sole purpose seems to be book promotion.

Be aware that the editors of the news outlet you are writing to may want to edit your letter. Make it clear that you want to sign off on any changes they make to your prose.

Finally, remember that letters are for amplification or argument, not for the correction of errors. If you are talking about an error in a report, you want to write a correction, not a letter to the editor.

One of the most successful (and prodigious) letter writers to the *Times* and to other publications, is Felicia Nimue Ackerman, a philosopher and bioethicist at Brown University. She writes so many terse, apt, and engaging letters that at times, at the *Times,* we worried we were running her correspondence too often.

I asked Professor Ackerman to share her secrets and she said she has none. She concentrates, she said, on "really obvious things like be concise and clear," advice she said would be helpful only to people who think they should be complex and verbose.[1] But as she described her letter writing, she did share some useful guidance.

First, she agreed that if you want to reply to an article or broadcast or posting, you should move quickly.

And think about your language, she added. "A lot of people nowadays talk mainly to people who agree with them" and who know their language, she said. "You're not doing that here. Scientists need to understand the need not to use technical language. They probably know that, but they have to keep it in mind." Her advice: run a draft of your letter by a nontechnical friend.

Ackerman said her letters are usually about things that annoy her. She said she wrote her first letter in response to an article about an effort to teach residents of nursing homes to write poetry. The teacher described the resulting poems as good, but Ackerman did not agree and, worse, she felt the instructor was patronizing his pupils.

Not surprisingly, her letters often draw negative

e-mail. Dealing with that requires a thick skin. Rejection is also an issue—she said only about a tenth of her letters see the light of day. But letter writing is rewarding, she said, in part because there is always a chance that a letter will encourage people to think about an issue in new ways. Ackerman also said she sometimes writes letters expressing ideas she believes many people hold but may be afraid to express. When people read them, she said, they "realize they are not alone."

Here are two examples of her letters published in the *New York Times.* In response to an article about brain damage in Alzheimer's disease, she wrote:

> It is hardly surprising that Alzheimer's disease can cause psychiatric symptoms stemming, at least partly, from brain damage. But it is important not to assume that this is the case whenever Alzheimer's patients think family members are out to harm them. Some families may be genuinely abusive toward Alzheimer's patients. Moreover, medicating a patient to make him, in his wife's view, "the same pleasant person he's always been," raises the issue of whether it is ethical to medicate a patient to benefit his caretaker.[2]

Whether you agree with Professor Ackerman or not, she has raised a compelling issue, and in only eighty-four words.

This one, even shorter, was written in response to an article about the sex researcher Alfred Kinsey:

> I was amused that Caleb Crain's article quoted Alfred C. Kinsey as saying "there are only three kinds of sexual abnormalities: abstinence, celibacy and delayed marriage" and also referred to Kinsey's "refusal to moralize about sex." Stigmatizing as "abnormal" a lack of sexual activity sounds pretty moralistic to me.[3]

Clear, cogent—and forty-nine words.

Will you achieve such miracles of expression? Maybe not. But if you keep these examples in mind you will improve your results.

OP-ED ESSAYS

The op-ed page—so called because it is the page opposite the editorial page—was invented at the *New York Times* as a vehicle for readers to express a wider range of opinion. Most newspapers that have op-ed pages do not limit their contents to one particular point of view. The point of the page is usually to enlarge the readers' perspectives by providing various opinions on many topics. Op-ed essays amplify or explain the news, describe its ramifications, implications, and relevance, or put previously unheard viewpoints before the readers.

I am amazed and disappointed that more researchers

don't take advantage of op-ed pages, or of the essay format generally, in a variety of journals and general-interest publications. For scientists, these platforms offer a rare chance to speak in their own voice to a lay audience, and a rare chance for the audience to hear (or read) that voice.

If you think you might like to write an op-ed piece, your first step is to read the publication you'd like to write for, to get an idea of its typical offerings. Get to know what you think works and what does not. Know the paper's rules for submission. Often these are printed on the op-ed page itself or on the publication's Web site. Once you have learned these rules, though, be prepared to break them. You might make an effective op-ed piece by annotating a chart or photograph, for example, instead of writing an essay. One of my students once made an op-ed piece by annotating a sketch by Darwin of a prototypical evolutionary family tree.

Have a point of view. Many newspaper editorial pages are afflicted with the "on the one hand this, on the other hand that" syndrome that makes for mind-numbing editorials. Don't contribute to this problem.

Be concise. A full column in a broadsheet newspaper contains less than 800 words. Except in extraordinary circumstances (and remember, an editor will be making that judgment) consider 800 words your upper limit. (The Gettysburg Address has 266.)

Competition for this space is intense. So explain quickly and clearly why your topic is important, why it is important *now,* and why readers should care what you think about it. Come across as a real person. Perhaps write in the first person.

In general, stick to exploring a single idea. At the *Times* we often cite William Safire, the one-time Nixon speechwriter and longtime *New York Times* op-ed columnist and wordsmith, whose theory is that one idea is a column, two is indecision, three is a trend (maybe), and four is a mishmash. Unless you are sure you have a trend, stick to one idea.

On op-ed pages, timeliness is hugely important. As a result, as with writing letters to the editor, if you seek success as an op-ed writer you should follow the news and be ready to pounce when an issue you care about moves into prominence. Send your submission by e-mail—but, again, don't send it as an attachment. Write it into your e-mail.

If your piece is accepted, be prepared to be edited. Be gracious. If you think an editing suggestion is ill-advised, ask what is motivating the editor to suggest it. She may have identified a problem you cannot see—

perhaps one that needs to be fixed but could be fixed another way.

Find out if the publication you have in mind will accept submissions that have also been sent to other publications. Try to find out when a decision has been made on your piece. I regret to confess that publishers of op-ed pages are notoriously unhelpful on this point. David Shipley, the op-ed editor at the *New York Times,* offers his guidance to would-be writers in a commentary available on the *Times* Web site.[4]

SeaWeb, a nonprofit organization that works to raise public awareness and understanding of marine and ocean conservation issues, also offers guidance for would-be opinion-ators. In general, the suggestions are similar to my own:

> Be timely. When your pet subject is in the news, move fast to submit your essay. How fast is fast? Within twenty-four hours. How can you possibly move that fast? By preparing ahead of time. At a minimum, know the news organization's rules for submission and so on.
>
> Consider your audience. Write clearly.
>
> If there are local angles—for example, if someone in the newspaper's circulation area is important in the debate, make that clear.

Provide graphics, maps, photos or other art to illustrate your piece.

Keep it simple—if possible, have one main point.

Use the active voice.

And if you don't succeed in getting published the first time, try again. You can do this quickly with e-mail. Don't think only of big publications. Accumulating a file of clips in small venues can help you break into the big time. Also, even if your publication of choice does not accept your essay, now the op-ed editor knows who you are and that you can write. He may come to you with an idea the next time news breaks in your field.

EDITORIAL BOARDS

Most large newspapers have an editorial board, a group of people whose job it is to determine what the newspaper's position will be on issues of the day and to write editorials on those matters. Often the boards reach out to experts in the community for guidance and information.

You can become one of their sources, particularly if you are knowledgeable about an issue important to the newspaper's community. If there is information you think the editorial board needs to have or a presentation you would like to make, write to the editor of the editorial pages and suggest it. If you are invited to meet with the editorial board, find out how much time you will have for your visit and prepare thoroughly but accordingly.

You may feel that it would be inappropriate for you to advance a particular policy position—but it can only be a good thing if the people writing editorials have the widest range of good information at their disposal. Surely you can take time to share *facts* with them. You might also suggest that they invite the reporter or reporters who cover your field to sit in on the meeting. We do this kind of thing regularly at the *Times,* and the meetings are almost always interesting and useful to the reporters who attend them.

12

Writing Books

Don't think about writing a book unless you really cannot help yourself.

I am not speaking here about the obligatory turning-the-thesis-into-a-book process that is a rite of academic passage. That is not book writing in the everyday meaning of the term.

I am talking about writing a book for ordinary lay readers, what people in the book business call a "trade" book, a book that comes out in hard cover, is aimed at the general public, sells in ordinary bookstores, appears again in a paperback edition, and maybe even finds a place in college courses. Writing this kind of book is very different from writing a paper or even a scholarly book.

If you write a book, of course you will have to marshal a large amount of data and organize it coherently. But you will also have to present your information in a way that moves the reader along with some kind of narrative drama. Also, while a scholarly paper works when its author presents new findings and ties them to other work, a book needs a point of view. While you probably don't want to produce a polemic, it is very hard to sustain any kind of narrative momentum if you are obsessed with giving equal time to all sides of an argument. And a book needs sweep. A narrow topic that works well for a scholarly (captive) audience may bomb with lay readers.

In short, it is not enough to produce a scholarly data dump. You must consider your readers and find a way to keep them engaged.

Book writing was never easy, but today's mass marketing techniques in book selling have made it harder than ever. The big chain bookstores want books that sell well, and they charge publishers for the prominent display space that can help turn a book into a best seller. (The same is true for some of the prominent display space given books on Amazon's Web site.) Most publishers won't take a chance and pay for that space without clear hopes of a big payoff. As a result, there is hardly any room anymore for what is called the "midlist" book—a book that sells steadily but not necessarily spectacularly—the kind of success a scientist or engineer is most likely to achieve as an author.

Also, as Susan Rabiner and Alfred Fortunato explain in their book *Thinking Like Your Editor*, for marketing reasons, bookstores tend to shelve books in only one place.[1] Getting your book placed where interested browsers will come across it can be difficult. The physicist Lawrence M. Krauss has long complained that his book *The Physics of Star Trek* is typically shelved with the heavy-duty science books and not also with Trekkie books.

Finally, consider this: At least 1,446,000 different titles were sold in the United States in 2006, according to Nelson BookScan.[2] Exactly 483 sold more than 100,000 copies. Four-fifths of them sold fewer than 100 copies.

If you have read this gloomy parade of facts and still think there's a book in you, you must ask yourself how much you (and your family) are prepared to sacrifice for it—not in money, necessarily, but in time, stress, and isolation. Ask yourself if you are willing to spend a lot of time on a project that may never see the light of day. Do you have that kind of discipline? Will your family support you or hinder you? Will they be upset if you spend early mornings, late evenings, or days off on this project? If your work is already solitary, do you want even more solitude?

Even if you can answer these questions in the affirmative, there are still some things to consider. You need to know who your book is for, why they will want to read it, and what else on the topic is already on the market. When prospective publishers ask who will buy your book, you will need to be prepared with an answer.

Don't think you are writing a book for the money. It is a truism among book authors that whatever your publisher offers you as an advance (amounts seem to be shrinking these days) is almost certainly all you will ever make from your book. In general, fewer than 10 percent of books earn back their advances. (That did not turn out to be the case with my first book, but that was largely because my advance was so small.) Consider whether the advance will cover the costs you will incur writing the book—travel, reference materials, permissions, copying, and so on.

Producing the manuscript is only part of the battle. In more and more publishing houses, editing seems to be turning into an optional activity and, believe me, everyone needs an editor. You may have to hire your own—another cost to consider. (Ask colleagues who have written books to recommend someone sane, competent, and, if possible, local.)

Also, once your book is published, you will have to promote it. If you are lucky, your publisher will arrange a media tour in which you travel to distant cities to appear on radio or television shows, give readings at bookstores, and so on. Would you enjoy this kind of thing? Can you spare the time for it? And if your publisher does not arrange this kind of adventure, will you take the time and spend the money to arrange it yourself? It is another truism among people who write books that publishers do not do nearly enough to promote them.

FINDING A PUBLISHER

But first you must find a publisher. One option is to approach an academic press, many of which are on the lookout for good technical subjects and good authors. Academic presses typically pay their authors less but keep their work in print longer, which is gratifying. Writing a book only to see it go quickly to the pulping pile is frustrating, to say the least.

If you want to publish with a trade press, or if you want help negotiating a better contract with any publisher, you will need an agent, someone to represent you in talks with publishers. Finding an agent can be difficult. One first step is to talk to colleagues who have published books. See who represented them and find out if the outcome was satisfactory. If it was, approach that agent and ask if she might take you on. Writing samples, a portfolio of clips and tapes or the like, can be helpful selling tools.

Your next step is to produce what is called a proposal. Ideally, your proposal describes your book, tells who its audience will be, and explains what they will draw from it. Your agent will send your proposal to publishers in the hope that one of them will see a book in it. In some ways, the proposal is the key to the success of the entire project, so be prepared to devote time and energy to it.

A typical proposal contains:

An overview, in which you explain what your book is about, what its argument will be.

An outline/table of contents.

One or more sample chapters.

Manuscript specifications (estimated number of pages, number of illustrations, whether they need to be in color, and so on).

A brief description of your audience. Tell your prospective publishers who your ideal reader is (and keep this reader in mind as you write).

A list of similar books that are already out there, and how your book will be different/better.

Your bio. Cast it in terms of what makes you the best person to write this book. It should be brief unless it is truly interesting.

Marketing opportunities, if any. Are you prominent in your field? Do you have any particular access to any particularly large/valuable audience segment? Don't be shy about mentioning these advantages.

If you have been on television or radio, include tapes. If you have received favorable reviews for earlier work, include those. Think in terms of materials your agent can use to sell your book to an editor, and an editor can use to sell it to the in-house committee that chooses which books the company will publish.

A successful proposal will demonstrate that you have convictions, the courage to advance and defend them, the ability to state them clearly and engagingly, and the passion to see the project through. If you are lucky, you will find an agent and your agent will end up fielding several offers for your book.

You may be tempted to choose a publishing house based on who offers you the most money. That approach is not necessarily as crass as it sounds. If a publishing house has invested a lot of money in a book, they will do more to promote it. But there are other considerations.

In his most recent memoir, *Avoid Boring People: Lessons from a Life in Science,* the biologist James D. Watson describes his experience writing *The Double Helix,* his elucidation of the structure of DNA. Among other things, he writes, the experience helped teach him that "a wise editor matters more than a big advance."[3] "Assuming you are not being insultingly low-balled, choosing a publisher on the basis of the advance is like choosing a house builder solely on the basis of the lowest bid," he writes.

Among other things, says Watson, an editor experienced in dealing with technical topics will realize that writing books about them may take longer than originally anticipated, and that illustrations, even color illustrations, may be essentials, not luxuries. (I lost this argument with the publishers of my first book, much to its detriment.)

It may be hard to discern, but look for clues about whether you will have a good working relationship with the person who will be your editor. You are looking for someone who will not let you get away with mere regurgitation of your academic prose but, at the same time, will not arm-twist you into science-lite. You may want to talk with some of the other researcher/authors the editor has worked with, to get a sense of her style.

The publisher's willingness to invest in producing and marketing your book is at least as important as the advance. Find out if the house has a record of success with your kind of book and a willingness to spend what it takes to make it a good book. In addition to editing and art production costs, for example, will the publisher agree to an index? The answer should be yes—insist on it. If your book is potentially useful in college courses, look for a house with experience selling to that market.

Even after you have settled on a publisher, your battles are not over. You may not like the suggested cover design or title, both of which can mean a great deal in terms of sales. You may have to insist that your publisher print an adequate number of copies in the first press run. According to Rabiner and Fortunato, an initial printing of 6,000 to 7,000 copies is typical, and that is obviously not very many.[4] I can say from experience that nothing is more heartbreaking (or more annoying) than hearing people say they wanted to buy your book but could not find it on the bookstore shelves.

FINDING A COAUTHOR

Many times publishers will suggest that researchers work with a coauthor, normally a journalist or science writer.

Think about whether you would enjoy this kind of collaboration and who an ideal collaborator might be. And think about whether it would make you crazy to have to consider the opinions of a coauthor.

If your agent or publisher suggests you work with a collaborator, consider the suggestion carefully. They may have reason to believe you need help with your prose. And if you decide to take this approach, how do you choose a good collaborator?

Your agent may already have someone in mind, someone who has worked with other authors, with good results. If not, start paying closer attention to journalists who cover your field in general-interest or even specialist publications, like *Nature* or *Science.* Suggest them to your agent or publisher.

My *New York Times* colleague Sandra Blakeslee has collaborated on books with several researchers with what she calls good results. How does she define a good result? "A good, readable book," she says.[5]

The most important thing to keep in mind, Blakeslee says, is something she once heard one of her scientific collaborators tell another would-be researcher/author: writing a book with a coauthor will not make the job easier, but it will make the book better.

"A lot of researchers think they can just turn it all over to the writer and the writer will magically read their minds and come up with a narrative and facts and a story that works," says Blakeslee. "That's just not true. You have to do a tremendous amount of work."

When Blakeslee works with a researcher-collaborator, she insists that the researchers periodically block out substantial amounts of time—one, two, or three days at a stretch—in which they can discuss one part of the book, a chapter, perhaps, and agree on what points it will cover and how they will be arranged. Then she writes up the material and sends it to the researcher, and they discuss it.

"Inevitably," she told me, "you take the material home and you write a draft chapter that you think is reflective of what they said, and they say 'Oh, I never said that' or 'You're putting words in my mouth.' And I have it on tape."

Inevitably, she went on, they'll eventually say, "Well, you're right, I did say that."

A good collaboration is like a good marriage, Blakeslee says. When one partner gets discouraged and thinks things are not going well, the other offers encouragement and support. "You want someone to buoy you up," she says. "That's the emotional seesaw of writing a book together."

Because of the need to work closely together, Blakeslee does not recommend working with a collaborator who lives more than 500 miles from you. Also, when she collaborates on a book, she insists on a fifty-fifty split of advances and royalties. "The researcher has to realize that the writer is not a serf or flunky or someone coming in to write your term paper," she says. "The writer is coming in with organizing skills, storytelling skills, interviewing skills, and can structure and pace a narrative."

The coastal scientist and author Orrin H. Pilkey, Jr., told me more or less the same thing about working with one of his collaborators. "I grew to appreciate and even envy his writing skills," he said. "We did have a few arguments about the way things were worded but I think our success was in each accepting the other as the most knowledgeable in certain areas."[6]

Some researchers already have excellent writing skills. If you think you may be one of them, Blakeslee suggests testing the idea by writing articles for nonscience publications. If your prose sails through with little difficulty, you may be right. But remember—it is not unusual for a scientist to win a large advance for a trade book only to turn in a manuscript the publisher deems unacceptable. If that happens to you, you may be obliged to return all or part of your advance.

Blakeslee is part of an eminent line of science journalists that began with her grandfather Howard Blakeslee and her father, Alton Blakeslee, and continues with her son, Matthew Blakeslee. "The scientist has to know about what my dad used to call 'the innocence of the reader,'" she says. He meant, she says, that readers are innocent of knowledge the researcher takes for granted. "If you expect that everybody knows what DNA methylization is—well, nobody knows what DNA methylization is," Blakeslee says. "But they could follow it with the help of a science writer."

SELLING YOUR BOOK

Once your book actually exists, it is time for you to promote it. You (or your agent) may have to nag your publisher for help—in the form of both buying good display space for the book in stores and paying your way to events around the country like book fairs, radio or television appearances, and book signings.

You can also promote your book yourself. These days, there are lots of ways to do it. A colleague's book became a best seller in part because he mentioned it on chat sites and elsewhere online. If you participate in e-mail forums or discussion groups, let other participants know about your book. Don't be shy about e-mailing family members, friends, scholarly collaborators—everyone you can think of—about your book. Mention your book in your e-mail signature so that everyone who receives an e-mail from you learns about your book.

If you have a blog or a Web site, these are obvious marketing venues. But if your Web site is full of family photos, field journals, and the like, consider setting up a separate page for your book. Also, check out your publisher's Web site and Web sites where books are sold, like Amazon, to be sure your book is described accurately and the description is accompanied by cover photos, sample chapters, and so on.

Offer to attend local community events and other occasions at which you can discuss, read from, and sell your book. When my first book was published, I kept one of the prepublication galleys as my "reading copy." I marked what I thought were particularly interesting or dramatic passages and turned to them when I was asked to discuss my book at a public meeting. You may want to select a few passages for their maximum dramatic effect.

Keep the book's publisher informed of these events, and give the company enough notice so that it can provide books for sale, possibly by working with a local bookstore. Be ready to sign books for people in your audience. If you see copies of your book on a bookstore shelf, approach the manager and offer to sign them.

Many stores like to display "signed copies." People love inscribed books—and they cannot be returned!

Many people, including me, are uncomfortable taking such a tooting-one's-own-horn approach. But unless you are prepared to abandon your book, abandon this reticence. Do you believe in your project? Then promote it.

It can be hard to tell what will work in promoting your book. Television helps books, but not as much as people may think. Oprah can turn any book into a smash hit, but even exposure on *60 Minutes* or *The Today Show* can be surprisingly unhelpful. Many people say the best place to appear is on National Public Radio, not because its audience is particularly huge—though it is impressive and growing—but because NPR listeners actually buy books.

The rise of self-publishing, especially with online firms that publish on demand, is changing the book business in ways we have yet to see play out. Self-published authors should realize, though, that at the moment it is hard to get book reviewers or radio and television bookers to pay much attention to self-published books.

Having said all this, I still think it's a great shame that more researchers do not write books for lay audiences, about their work or even about themselves. I have always thought of science and engineering research as a quest—the accepting of some technical challenge and the work of years to overcome it. I view this enterprise as romantic. Telling the story of this work is a way of drawing ordinary people into its fascination.

As Georgina Ferry put it in an essay in *Nature,* scientific biography is a way of leading readers to be more "alive to the cultural significance of scientists."[7] But few scientists do that kind of writing. Ferry, a biographer of the chemist and Nobel laureate Max Perutz, argues that scientists' reluctance to embrace biography or autobiography maintains a "gulf" between the world of research and the world outside.

"Scientists publish their work in places where only other scientists will read it, in language that only other scientists understand," Ferry writes. They contend that writing about themselves will attract the derision of their research peers (the fate that initially greeted Watson's book *The Double Helix*). Others reject tales of scientific rivalries and personalities as little more than gossip. Writes Ferry: "Some argue that individuals are irrelevant to the progress of science; anyone could have discovered the double helix but only Leonardo da Vinci could have painted the Mona Lisa."[8]

But, Ferry points out, research is carried out "by real people" with real, individual attributes. "Telling their stories transforms the stereotype of the scientist into vivid individuality," she writes. A truthful biography or autobiography tells of research collaborations and rivalries, pitfalls and triumphs, family pressures and support, and so on.

These are the kinds of stories that people like to read. And, done well, they can educate people about what it means to engage in research. It will be grand when the research community accepts, as Ferry puts it, that telling these stories "depends simply on the willingness of scientists to appear as individuals—and of their colleagues to applaud their doing so."[9]

Writing a book, seeing it to publication, and then promoting it can be arduous and frustrating. But the rewards are great. Your book is out there. If it stays in print, it is out there for a long time. And it will live even longer in libraries and second-hand book sales, which are much more practical now than they used to be. With luck, your book will be a useful guide, leading your readers to better understanding and possibly even better actions.

13

On the Witness Stand

"As society becomes more dependent for its well-being upon scientifically complex technology, we find that this technology increasingly underlies legal issues of importance to us all." That's what Supreme Court Justice Stephen Breyer wrote in a decision in *General Electric Co. v. Joiner,* a case involving the admissibility in court of expert testimony.[1] Examples of these technology-related issues are all around us: microwave radiation, mercury in vaccines, asbestos exposure, and, in the environment, everything from the Endangered Species Act to climate change—these and other topics regularly turn up in the nation's courtrooms.

But science in the courtroom is problem plagued. Many people believe the law does a poor job of handling technical evidence. Worse, they believe law and research will never mesh well because, while in theory each is engaged in a search for truth, they approach the search in fundamentally different ways.

Science is an unending search for explanations in which the questioning process prevails and all answers are temporary—valid but only provisionally, pending a successful challenge by new findings. Scientists and engineers start with questions and look for answers. They collect all the information they can and then draw conclusions. It would be unethical for them to ignore findings that don't fit their hypotheses.

By contrast, the law works on the premise that the best way to learn the truth in a dispute is to have each side advance the strongest arguments it can to make its case. Lawyers on each side know their desired outcome at the outset, and gather arguments and evidence to support it. They are not necessarily obliged to produce evidence that hurts their case; that's the other side's job.

As the National Science Foundation (NSF) put it in a report on the subject, the cutting edge of scientific knowledge is always moving, sometimes rapidly.[2] Meanwhile, legitimate scientific or technical differences can prevent a consensus from emerging. And some issues are inherently uncertain. As the NSF says in its report, "Many phenomena behave stochastically and simply cannot be known in other than probabilistic terms." Our legal system is not designed to handle that kind of uncertainty.[3]

Plus, courts deal only with what is presented to them. If information is not introduced in testimony, it may as well not exist. If evidence is introduced without contradiction, it stands unchallenged.

And of course neither the standard of truth—a judge or jury's subjective judgment that "a preponderance of the evidence" is on one side or the other—matches well with the researcher's standard of "statistical significance," which indicates that the odds that a finding happened by chance are less than one in twenty.

But in some ways, researchers are given special privileges in court. Most of the time, for example, witnesses in a trial may testify only to what they personally saw, heard, tasted, or otherwise experienced through their own senses. By contrast, scientists or engineers, as expert witnesses, are brought in to assist the judge or jury—the "trier of fact," in legal parlance—with explanations and interpretations of the facts. So they have leeway other witnesses do not enjoy. They can speculate, offer opinions, and so on.

Who should be allowed to don the robe of the expert,

and what kind of testimony should that person be allowed to offer? The questions are important because in many instances expert technical testimony can make or break the case. For example, if the issue is whether a pharmaceutical company is liable for adverse side effects of a drug, or a shopping center for degradation of a wetland, the case may be won or lost on testimony about how the drug works in the body or how the design of a parking lot affects runoff.

Often, judges and juries find themselves facing issues in which the science is unclear. Often, in fact, issues do not attract much research firepower until they have moved into the courts. That's what happened with litigation over breast implants, dioxin pollution, and other issues.

While scientific knowledge can advance quickly, legitimate scientific or technical differences in research can prevent a consensus from emerging. Some issues are also inherently uncertain. Our legal system is not designed to sort out this kind of uncertainty.

In recent years, guidelines for the admissibility of technical evidence have evolved. For years, the requirement was "general acceptance" by the scientific or technical community, a standard that a federal appeals court enunciated in 1923 in *Frye v. United States,* a case involving the admissibility of the results of lie detector tests.[4]

Though the court noted that it was difficult to say "just when a scientific principle or discovery crosses the line between the experimental and demonstrable stages," it ruled that polygraph technology had yet to win wide acceptance, and barred it from the courtroom.

But in 1973 Congress codified new federal rules of evidence. Under these rules, the admissibility test became not wide acceptance but whether "scientific, technical or other specialized knowledge will assist the trier of fact to determine a fact in issue," and whether the witness is qualified to speak on the issue by virtue of knowledge, training, or expertise.[5] In other words, the focus shifted from the quality of the testimony to the qualities of the witness.

Critics of this standard say it brought a flood of junk or fringe science into the nation's courtrooms, as people with good (or seemingly good) credentials took the stand before juries unable to tell the difference between credible scientific or technical claims and those with only the aroma of respectability.

The breast implant case offers a good example. Though there was never any convincing epidemiological evidence that implants caused systemic disease, women filed lawsuits charging they had contracted everything from "malaise" to multiple sclerosis as a result of having them. By the time an expert panel weighed in, years later, an implant manufacturer had gone out of business and plaintiffs had collectively won billions of dollars. The panel found that while implants might leak or produce localized scarring, they were blameless so far as systemic disease is concerned.[6]

On the other hand, lawyers for people seeking damage awards would say industry is quick to apply the label "junk science" to any opinion it does not like.

Finally, in 1993, the Supreme Court spoke on the admissibility issue, in a case involving allegations that the antinausea drug Bendectin caused birth defects. (It did not, but testimony that amounted to little more than post hoc, ergo propter hoc—"after this, therefore on account of it"—resulted in so many jury verdicts against the drug that its maker, Merrell Dow, pulled it from the market.) In the case, *Daubert v. Merrell Dow Pharmaceuticals,* the Court held that judges had the obligation and the capacity to assess whether proposed testimony was scientifically reliable and relevant. And it offered them some guidelines for doing so, such as considering whether the theory or technique in question has been tested, or can be (the decision actually cites the work of Karl Popper and his idea that a scientific idea is one that can be tested in the real world); whether the theory or technique has been reported in the scientific literature, subject to peer review; and whether there are standard error rates.[7]

In *Joiner,* a subsequent decision in 1997, the Court noted that conclusions and methods are inextricably related. Therefore, according to the decision, "a court may conclude that there is simply too great an analytical gap between the data and the opinion proferred."[8]

This line of opinions, while reasonable, has left a lot of judges worried. Few judges have advanced training—or indeed any training at all—in science. How are they to decide even these preliminary questions of science in the cases that come before them?

Acting on their own, one group of judges organized a conference at which experts could tell them about DNA. The federal judiciary also has a program of training sessions for judges. Sometimes courts appoint scientific masters assigned to evaluate both sides of an issue. In the breast implant case, for example, a judge eventually had experts examine available evidence on four points: epidemiology, immunology, rheumatology, and toxicology. The experts, in turn, eventually produced reports finding no connection between the implants and systemic disease. The judge had appointed a panel of six special masters to choose the experts.

Some who follow these issues have proposed establishing special science courts, where questions of scientific evidence could be ruled on. But, according to Sheila Jasanoff of Harvard, a scholar of law and the history of science, American lawyers don't like the idea of expert scientific witnesses paid by the courts because they fear the power they will have over judges and juries.[9] Many lawyers also object to, in effect, turning control of their cases over to a panel of experts.

In any event, nothing has so far done away with the need for scientists and engineers on the witness stand. If you are asked to be one of them, you have serious points to ponder.

For one thing, agreeing to testify for one side in a case invites potentially unpleasant scrutiny from the other. Robert J. (Skip) Livingston, a biologist at Florida State University whose testimony has been instrumental in important conservation decisions, describes the experiences as "a whole lot of hostile people trying to nail you to the wall."[10]

In *Tainted Truth*, Cynthia Crossen describes the witness stand as "the slaughterhouse of reputations."[11]

If you do agree to serve as an expert witness, you risk embarrassment or even (in theory) a perjury charge if you change your testimony for any reason. And in at least one case I know about, a coastal geologist was himself sued after testifying that a particular bluff on the

Oregon coast was a dangerous site for a condo development because of the bluff's risk of slumping. The suit was eventually dropped, but it caused the scientist many uneasy moments. (The condos were approved and construction was begun but, just as the scientist had predicted on the stand, the bluff soon slumped beneath it and the project was never completed.)

This kind of legal action is an example of "strategic litigation against public participation"—a SLAPP suit. Witnesses cannot be sued for things they say on the stand, but people intent on discouraging testimony by scientists, engineers, and others may track their every utterance outside the courtroom and sue them for defamation or other torts.

As in the Oregon case, SLAPP suits usually fail, at least in court. But because defending against a SLAPP suit requires time, energy, and money, the suits can succeed in the public arena simply by discouraging potential witnesses from speaking out. Also, they can delay resolution of the underlying issue, which often cannot go forward until the SLAPP suit has been dealt with.

Some states are taking action to discourage SLAPP suits or minimize their effects. According to the First Amendment Project, in 1993 California changed its Code of Civil Procedure to require judges to dismiss SLAPP suits unless the plaintiff can demonstrate a "probability of winning." If he cannot, he must pay the target of his suit the costs of defending against it.

You can protect yourself, to an extent, by making sure that everything you say is absolutely accurate. Be ready with studies, reports, and the like to back up your statements. But because this kind of litigation is launched not on the merits but as a terror tactic, substantive issues may not be the point.

If you are sued, obtain advice from a lawyer quickly. Note that you will have a limited amount of time to reply to the suit, so don't delay. If you have homeowner's insurance, ask your carrier or agent if it provides any coverage for you in the case that you are sued.[12]

So there you are—challenged on the witness stand, subjected to nasty cross-examination, maybe even sued. And your colleagues may even accuse you of grandstanding. Why bother?

Well, my first answer would be because the rest of us are relying on your expertise. But there's another reason. While publication in the scholarly literature is a great thing, relatively few people read the scholarly literature. Your information may take a long time to percolate into the world in a useful way. By contrast, a court verdict for your side resonates immediately. Successful litigation can do more good, faster, to help spread your word than dozens of scholarly publications.

Still—testifying in court is a big commitment of time and energy. Don't even think about it unless you are confident the case has merit and the lawyers on your side are competent. And then ask yourself if you are the best person to testify. If the truthful answer is no, decline the honor. No matter what, inform the lawyers on your side about any gaps in your training or other problems your opponents could use to discredit your testimony.

If you will be paid, consider whether conflict of interest is likely to be a problem for you—whether you personally have anything hanging on the outcome of the litigation. That could make you an ineffective witness.

Also consider whether participating in the trial as an expert witness for one side might cause you a conflict-of-interest problem in the future. For example, Peter Shelley, a lawyer who litigated cases involving pollution in Boston Harbor, says it was hard to find an expert witness who would testify because they all hoped eventually to be paid to work on the harbor cleanup. [13]

Anticipate what data the other side will use to make its case and know in advance how to undermine them. Figure out how to do that in language that is technically accurate but simple enough for a jury to follow. As Livingston put it, "A lot of jurors are bored with this stuff. You have to boil it down to the lower limits."[14] Is this possible? Yes, always. But ask yourself honestly if you are prepared to invest the time and effort.

Should you be paid for serving as an expert witness? Well . . . that depends. Plenty of litigants will have a substantial budget for witnesses, and some jurisdictions have scientific witness funds that intervenors in important cases can have access to. In fact, some researchers have turned courtroom testimony into a lucrative sideline. But your effectiveness can be diminished if you are perceived as a technical mercenary.

Suppose you have considered all these factors and still decide to testify—what happens next? In trials on television and in the movies, people routinely make stunning announcements from the witness stand, changing the course of the case. In real life, this kind of thing is highly unlikely. In real life most of the important trial action occurs before the case reaches a courtroom, if it ever does. There are so few surprises because each side in a case has a right to know, ahead of time, who their opponents intend to call as witnesses and what kind of testimony those witnesses will offer—allowing them to collect material ahead of time to rebut the testimony.

The two sides learn about potential testimony by questioning each other's witnesses in what's called the discovery process. The result is that the court proceedings often turn into a one-time-only production of a drama already rehearsed behind the scenes in question-and-answer sessions called depositions. In a deposition, a lawyer for the other side will ask you about your prospective testimony. These proceedings will almost certainly be recorded, and you answer under the penalties of perjury.

Realize going in that if they have prepared thoroughly, lawyers for the other side will have read every book or article or conference paper you have ever written, looking for weak spots, contradictions, anything that would challenge your credibility. If these weaknesses exist, make sure your lawyers know about them, and discuss how you will deal with them in deposition.

Once I was questioned in a deposition, part of the discovery process in connection with a libel suit brought against the *New York Times.* I learned some valuable lessons.

"You cannot win the case in deposition—there's no judge there," one of our lawyers, George Freeman, told me. "You can only lose the case," he said, by saying something that gives the other side a weapon it can use to defeat you. So, Freeman said, it is important to pause before you answer any question put to you in a deposition. This will allow you to think through exactly what you want your answer to be. More important, it will allow your lawyer to object if for some reason you ought not to answer the question in the first place.

Perhaps, for example, you're asked a trick question of the "when did you stop beating your wife?" variety. Your lawyer can protect you from this kind of questioning, but only if you allow him time to intervene with an objection.

Outright trickery is rare, lawyers I have talked to say, but it is typical for lawyers from the other side to present themselves as friendly and encourage you to be chatty. Resist this impulse. You are not there to make friends.

Resist also the urge to appear knowledgeable. Answer only the questions asked, and don't volunteer other information. Think of yourself not in terms of the handful of other experts in your subspecialty, but in terms of how you can explain things to the general public (the judge and the jury). At the same time, note that what you say in discovery may be all you will be allowed to talk about at trial. Opportunities to introduce other subjects in testimony may be limited. If there is an important issue not being raised, tell your lawyers.

Don't allow yourself to be badgered. If you don't know the answer to a question, or cannot remember, just say so. Don't give an answer unless you are sure of it.

If you don't understand a question, say so. Don't answer the question you *think* the opposing counsel is asking. If, when you are later on the stand, the question is explained more fully and you answer it differently, your veracity might come into question.

Don't be jocular. This little bit of advice seemed so self-evident I was surprised to receive it. I can only assume the lawyers have learned by bitter experience to offer it.

You may be asked something like "Have you talked to anyone about what you are going to be deposed about?" Just tell the truth. There is nothing unethical in preparing for a deposition, or in getting your lawyer's advice on this process.

Don't argue with anybody's lawyer, and never, ever, argue with the judge.

Finally, remember that the standards that apply in the lab differ from the standards of the courtroom. In research, it is ethically wrong to discard uncomfortable data. But it is not considered unethical to leave information out of a legal argument. Yes-or-no questions are fair in court. Prepare with your lawyers to answer them.

It's not that lawyers work in a different ethical universe from yours (they say) but that they are working with different rules. Are you uncomfortable with this? Ask that question ahead of time, and bow out if you are.

14

Making Policy

Many researchers believe that when political disputes hang on questions of science, government officials, regulatory agencies, and ordinary people would do the right thing, if only they knew the facts. That is, they believe that if everyone knew what *they* knew, everyone would think like they think, and if policymakers make bad decisions, it must be because they are ignorant of science or technology.

That's just not the case.

As Congressman Sherwood Boehlert of New York told a meeting of the American Association for the Advancement of Science in 2007, "Science has to inform policy making, but it isn't determinative. Pretending that science is going to settle a dispute that is really about values or money or anything else just leads to muddled thinking and distorted debates that are damaging to both science and policy in the long run."[1] In other words, technology- and science-related policy, like all policy, is a manifestation of social aspirations and values expressed through political action.

Congressman Boehlert, a Republican and one of the last of the party's northeastern liberals, retired in 2007 from the House of Representatives, where he headed the Science Committee. Scientists should absorb his wisdom if they hope to be effective in the policy process.

Sometimes the "values" issue at stake is obvious. For example, your views on research involving human embryonic stem cells or fetal tissue probably depend on your views on the moral status of the embryo. And sometimes the values involved are less obvious. Climate change can be thought of as a values question, in that it hangs on whether this generation has the right to leave behind a degraded planet for the next generation.

But often policymakers don't want to confront these knotty values issues directly, so they disguise their arguments as disputes about scientific or technological uncertainty. Instead of arguing, say, that the destruction of human embryos is always morally wrong or that immediate economic growth is always more important than long-term ecological health, they will argue that nothing can be achieved using embryonic stem cells that cannot be achieved with adult stem cells, or that there is no persuasive evidence that human actions are adversely affecting the atmosphere.

Science and technology seem to be powerful weapons in this kind of argument—not because they can answer the questions but because of their credibility. In survey after survey, Americans list scientists among the people they most respect. As Boehlert put it, "In our highly polarized political environment, describing your position as the only scientifically valid stance is perhaps the only remaining way to seem more 'pure,' more convincing, more above the fray than whomever you're sparring with."[2]

One of Boehlert's favorite examples, he said in his AAAS speech, was a 1997 Clinton administration proposal to toughen clean air standards for ground-level ozone. Boehlert supported the move, he recalled in his speech, because it was "pretty clear" that increased ground-level ozone led inexorably to increased hospital admissions. But where to draw the line? How many additional hospital admissions are acceptable?

"This was a simple, direct and horrifying question, and no one wanted to go near it," Boehlert recalled. So the debate degenerated into one about what kind of ozone decision would be "scientific." Of course, there was no such thing.

There was a time when the government's own science establishment might have been able to weigh in. In the years after World War II—won, many believed, by radar and the atomic bomb—science and engineering had abundant respect in government. The launch of Sputnik in 1957 and the ensuing space race only raised the ante.

In the years since the end of the cold war, however, the influence of scientists has waned. The Office of Science and Technology Policy—which serves as science adviser to the president—was downgraded in the administration of George W. Bush. (Researchers have high hopes for it in the Obama administration.) After the "Republican revolution" of 1994, Congress abolished the Office of Technology Assessment, a much-beloved congressional institution. And though the House still has its Science Committee and, more important, the Science Committee's staff, there is no direct equivalent in the Senate.

That leaves Congress to rely on the Congressional Research Service, a think tank that provides information to Congress, and the members of the National Academy of Sciences who, while highly respected, move with such deliberation that a six-month turnaround time between a request for information and the production of a report is regarded as spectacularly speedy. Also, when the National Research Council, the academy's research arm, appoints a panel, its members typically strive so diligently for consensus that their reports often lack punch.

Additionally, because public officials will rarely support anything, however worthy, that will cause big trouble for a powerful constituent, they may unconsciously absorb only the information that backs the choice that they know, as a practical matter, they must make. Politicians also realize that while people routinely list some technology based issues—environmental protection, for example—as matters they are highly concerned about, they almost never turn science issues into "voting issues." That's why people like Daniel Sarewitz, director of the Consortium for Science, Policy and Outcomes at Arizona State University, calls policymaking "an unhealthy dynamic" that researchers should steer clear of.[3]

Again and again scientists tell me, in effect, that they see his point, and they worry that even offering information can somehow compromise one's scientific objectivity. Many are contemptuous of scientists who work in regulatory agencies. As one coastal engineer once put it to me, "Regulation is the revenge of the C student."

The upshot, as Daniel Greenberg puts it in his book

Science, Money, and Politics, is that "scientists today are largely missing in action from American politics and public affairs beyond the clear boundaries of science," by which he means the boundaries of science funding. He adds, "Advanced scientific training is astonishingly rare among elected officials at all layers of government."[4]

Regardless of what other pressures may weigh on them, when policymakers must act on technology-related issues, it is important that they at least have access to the best information available. That can come only from scientists and engineers. As Boehlert told his AAAS audience, "Scientists should participate actively, even avidly, in policy debates. Indeed, both as educated citizens and as professionals with relevant knowledge—not to mention as beneficiaries of public support—scientists ought to feel obligated to contribute to policy making—in their communities, in their states, in the nation, and even in the wider world."[5]

But as Martin Rees, the eminent British cosmologist and astrophysicist, put it in an essay in the *New York Review of Books,* "Policy decisions—whether about energy, GM technology, mind-enhancing drugs, or whatever—are never solely scientific: strategic, economic, social and ethical ramifications enter as well. And here scientists have no special credentials."[6]

Scientists should not be indifferent to the fate of their

ideas in the wider world, Rees said. They should promote their benign use and campaign against their misuse. "And they should be prepared to engage in public debate and discussion."[7]

LEGITIMATE ADVOCACY

But what kind of action is legitimate? Should researchers limit themselves to offering policymakers "just the facts"? Or should they make policy recommendations or even advocate for particular policies? There are few established professional standards.

Few people would question advocacy of increased research funding. When the Bush administration proposed limiting access to the libraries of the Environmental Protection Agency, the researchers who protested, successfully, were praised for their efforts. But when Robert Jarvik, MD, inventor of the artificial heart, appeared in television commercials for the cholesterol-lowering drug Lipitor, he came under intense criticism—in part because he was not a cardiologist, or even a practicing physician.

You might regard those examples of advocacy—speaking out for research as an enterprise and taking money to advance a particular scientific position or product—as two ends of a spectrum of activity. There is much in the middle. Is accepting money to endorse a product always unacceptable advocacy? When health researchers reported a steep rise in civilian death rates in Iraq since the U.S.-led invasion, was that advocacy? How about when they published their results just before an election?

It is difficult to answer these questions. Science's long-standing reluctance to engage in public issues has left it, on the whole, without well-established, clearly understood guidelines for how scientists should behave in the public arena. Guidelines that do exist are not necessarily well known, much discussed, or vigorously enforced. And I hesitate to suggest that you discuss the issue with senior faculty members or other gray eminences at your institution, for fear they will simply instruct you to keep your mouth shut and your head down. So I will try to be satisfied with offering some advice I have heard activist scientists of one kind or another offer to their colleagues.

One unobjectionable approach, adopted by many scientific and engineering organizations, is to issue formal position statements on matters being argued in public. For example, in 2008 the National Academy of Sciences issued a book arguing that the theory of evolution is the foundation of modern biology and medicine, that there is no credible challenge to it, and that accepting evolution does not imply rejection of religion. The journal

Nature, in an editorial, greeted the book with "three cheers" and urged researchers to "spread the word."[8]

In December 2007, in another example, the American Geophysical Union issued two statements. One, a revision of earlier statements, emphasized the importance of teaching the theory of evolution and the geological history of the earth as "foundations of science." The other, a joint effort with the Seismological Society of America, was in effect an endorsement of international nuclear test ban treaties—the statement asserted it would be practically impossible for a cheating nation to test nuclear weapons without being detected.[9]

It's great when professional scientific organizations take this kind of a stance, and I wish more of them would do it more regularly. For example, when a microbiologist asked me if it would be helpful if his society put a statement on its Web site on the plausible evolution of the bacterial flagellum, a perennial creationist red herring, I could only agree. Statements by professional societies can provide very useful backing for people trying to fight the forces of ignorance in our society.

But statements can go only so far.

Nature, in its editorial praising the National Academy of Sciences' book, said it was not enough for science groups to take a position. They should also, the editors said, "take every opportunity to promote it."

Two earth scientists, Timothy Dixon and Roy Dokka, made a similar point with unusual force in an essay in the American Geophysical Union's weekly publication *Eos*. Their title tells it all: "Earth Scientists and Public Policy: Have We Failed New Orleans?" It was abundantly clear before Katrina that the city was in extreme danger, they wrote: "The area is a delta; it is low lying and getting lower all the time; it will flood again; lowest elevations are most at risk; and levees must be periodically augmented to keep up with subsidence, sea level rise, and future storms. Why were earth scientists not able to deliver this simple message?"[10]

They answer this question in a way that applies to scientists in many disciplines. When earth scientists are asked for policy advice, they said, "We tend to speak to our peers rather than to the public"—in cautious, jargon-filled, caveat-ridden language. "Obvious, common-sense statements are often avoided," they added. "Also, public policy statements tend to come from committees, where clarity is sacrificed for consensus."[11]

"Instead of providing a simple picture of how low elevation, unrelenting subsidence and sediment starvation set the stage for coastal flooding," Dixon and Dokka wrote, "the Earth science community emphasized the complexity of the problem and disagreements among scientists."

This reluctance to cut to the chase is frustrating to policymakers, to say the least.

TALKING TO POLITICIANS

So suppose you accept the challenges of citizenship and decide to speak up a bit more. Here are some practical tips. I use Congress as my example, but my advice applies (more or less) across the board. I have absorbed it by talking with people who work as elected officials and by watching those who lobby them successfully—or unsuccessfully—on matters of science. I think it applies across the political spectrum.

You must differentiate between professional opinions, personal points of view, and the positions of your institution. And you should level with your audience. If there is substantial, reasonable opinion at odds with yours, say so. This frankness protects your credibility and, in the end, strengthens your message. It is not necessary for someone to be an evolutionary biologist to advocate for the teaching of evolution, but if you are recommending a particular policy for standards of offshore salmon farming, for example, it's best if you speak from knowledge and experience.

Avoid personal gain. Again, your credibility is at stake. Note, however, that there are situations in which it will not be possible to separate your personal gain from your advocacy. If you are a stem cell researcher, for example, advocating for the financing of stem cell research relates directly to your personal fortunes.

"You have to appeal to the interests of the audience that you're dealing with," Congressman Boehlert told the journal *Science* in an interview in 2006. "To talk about some great advance in pure scientific terms isn't enough. What does it do to strengthen the economy or enhance competitiveness or provide more jobs?"[12]

"We don't have time for tutorials," Boehlert added. "They need to get right to the point: this is why it's important. I know there are a lot of competing interests, but here's why we should be at the head of the line. And here's what it means for society." Not everyone has natural instincts for this kind of conversation, he acknowledged, adding, "People who are good at it have to train their colleagues."[13]

To get your message across, take the following steps:

> First, get a good referral. If you know someone who will support you—and to whom the official you need to talk to must pay attention—get that person to put in a word on your behalf.
>
> Find out as much as you can about the official you want to talk to—his politics, his district and its con-

cerns. For Congress, the *Almanac of American Politics* is an excellent source. If your institution has a government relations staff, maybe they can help you. Or you can consult your local newspaper's archives. Does the public official have a staff? Find out who is on this staff, what issues they deal with, and whom you should approach.

Present your views in a professional manner, whether by e-mail, on paper, or in person. Your dress, manner, language, and so on should be appropriate to your audience. As Linda Michaluk of the Association of Professional Biologists of British Columbia put it at the 2008 AAAS conference, "Sometimes the suits need to see a suit to hear what is being said." At the same time, said Professor Michaluk, "when I show up in that little black dress and pumps and I am speaking to the Raging Grannies, that isn't going to work."[14]

Everyone resents being patronized, but this resentment is acute among public officials. When you draft a letter or e-mail or plan a phone call or visit, run your message past a friend or colleague who will warn you if you are coming across as arrogant.

If you have an appointment with a policymaker or politician, be on time. But be aware that the person or people you are meeting may run late, not necessarily through thoughtlessness but because of the vagaries of committee meetings, floor votes, and so on. Consider meeting congressional staffers in their district or state office (though issue-oriented staffers are more likely to be based in Washington).

Keep your message simple, focused, and short. Provide context, but only enough to make yourself clear.

Encourage questions, and if someone asks a question you cannot answer, don't fake it. Say "I don't know" when appropriate and offer to provide the needed information later. But as the old saying goes, don't let your mouth write checks your ass can't cash. If you promise to provide additional information, memos, or the like, be prepared to produce them, and fast.

Think about the best time to make your pitch. You don't want to speak too soon, before an issue has really registered on the official's mental radar, but you don't want to wait until things are so far along it may be hard to modify them. For example, while it is theoretically possible to modify legislation once it reaches the floor of the House or Senate, or even when it reaches a House-Senate conference committee, it is much easier to have an impact if you present your information while a bill is being drafted—during hearings on the matter, for example.

Beware of overhyping the importance or certainty of the information you want to convey. Be up-front about uncertainties, data inconsistencies, blank spots in the record, and so on. Don't base a strong pitch on a theory or finding if there is a chance it may soon come apart. On the other hand, don't undercut your message unnecessarily.

Here's an especially important bit of advice: don't be someone who can only point to problems. Be ready with a possible remedy or, if that makes you too uncomfortable, a menu of possible approaches to the issue. Describe their strong and weak points. Show how the actions you describe affect the matter at issue. Demonstrate that you can *help* the policymaker. Don't just make his life more complicated or difficult.

After your meeting, thank the official for his time, preferably in a handwritten note, perhaps briefly reviewing your discussion, noting what additional information you will provide, and so on. Remember, this is a thank you. It is not an opportunity to restate grievances or rehash disagreements.

Invite the official to attend local meetings of scientific societies or to visit your lab or local factories using your technology. Make the same invitations to the politician's staff.

That's what Vernon J. Ehlers did in the 1970s, when he was teaching physics at Calvin College in Grand Rapids, Michigan. Having met his congressman, Gerald R. Ford, he says, "I thought, well, I'll send him a letter and offer help."[15] Ehlers offered to put together a committee of scientists who could meet with Ford to discuss science aspects of policy questions. To Ehlers's surprise, Ford was delighted with the suggestion, and Ehlers ended up assembling a group of scientists from the district, some Republicans and some Democrats, who met with Ford three or four times a year, more often if a particular scientific or technical issue was on the boil.

This work had an unexpected payoff for Ehlers, a Republican. Today, he represents the district—one of a small number of scientists in the Congress. "Ford genuinely seemed to enjoy these meetings," Congressman Ehlers recalled. "I told him, 'You never look at your watch during the meetings.' And he said 'Vern, I meet with people all day long and you are the only people I meet with who are trying to give me something, not ask me for something.'" That is the kind of relationship you should try to cultivate.

TESTIMONY

Shortly after Hurricane Katrina struck the Gulf Coast, Rob Young, now head of the Program for the Study of Developed Shorelines at Western Carolina University, invited me to accompany him on a flyover of the devastated coast. I wrote about what I saw in the science section of the *Times*.

As a result of the article, Young was invited to testify before Congress, in a hearing on coastal policy. While this kind of thing can turn into mere political theater, it can also be a valuable opportunity to get important information on the record, not just in the record of the hearing itself, but in the expert comments considered by those writing, enforcing, and litigating laws. If you are invited to testify before a congressional committee, or to appear before a town council meeting or the like, recognize that you have been given an opportunity to do some good, to put your information and ideas out there in a venue potentially far more influential than a professional meeting or scientific journal.

Should you receive such an invitation:

Consult ahead of time with the staff people organizing the hearing or presentation. Understand whether its purpose is to put information on the public record, inform committee members, challenge prevailing ideology, or, as is sometimes the case, force legislators to confront an issue by bringing it to the public's attention.

Know who the other witnesses will be, and understand how your testimony will or will not mesh with theirs. (Rarely, a witness will be invited to play the role of sacrificial lamb—to be sacrificed at the witness table. You can avoid this fate by inquiring thoroughly ahead of time about the circumstances of the hearing.)

Meet administrative requirements. For example, be prepared to speak for the time allotted and no more. (In my experience, it takes about eight minutes to speak a thousand words. Time yourself and see how you do.) Arrive at the hearing room at least thirty minutes early. Submit written testimony at least forty-eight hours ahead of time, except in highly unusual circumstances. Often witnesses are allowed to provide a relatively large amount of written testimony—if you do this, be prepared with a terse summary.

As always, write with simplicity, brevity, and clarity in terms for the lay person. Use simple examples. If you have to talk about data, see if you can provide charts, not just for the written testimony but large ones for use in the hearing room. William G. Wells, Jr., in his book *Working with Congress,* advises thinking of the proceedings as theater in which you are the star.[16] Force the audience members to listen to you!

If your testimony relates to a measure under consideration, make sure your remarks will be included in the "legislative report" that accompanies it. Then they can be given official consideration in subsequent litigation, if any.

Afterward, read the transcript to be sure there are no mistakes. If you find any, try to get them fixed as soon as possible.

Finally, after the hearing, ask committee staff members, or the staffers of the member who invited you to testify, how they thought your testimony went. You may learn something that will improve your performance next time.

Another way to convey information to Congress is to make a presentation in the Capitol open to anyone who wants to listen. For example, in 2006, scientists at the Harvard University Center for Health and the Global Environment, with the backing of senators Richard Lugar and Barack Obama, organized a briefing session titled How Science Works and How Can Science Most Effectively Inform Policy? With the senators' help, the center reserved a room in the basement of the Capitol and recruited speakers—including Donald Kennedy, then the editor of *Science,* and Harvey V. Fineberg, president of the Institute of Medicine—to describe how scientists do their work, test their findings, report them to their peers, and so on.

The organizers of this event, led by Eric Chivian, director of the Center for Health and the Global Environment, said they were motivated by the continuing debate over the teaching of evolution, and the evident inability of many Americans to differentiate between an idea based on science and one based on the supernatural. Only a few members of Congress attended the session, but the room was packed with people who may almost be more important in issues of this nature—congressional staff.

Some audience members were young researchers interested in learning how science or science-related policies are made. They were working in the Capitol as part of a fellowship program of the AAAS that places scientists and engineers in federal agencies and congressional offices, where they work for a year. The hope is that not only will they demonstrate to their Capitol Hill colleagues how researchers view the world but also that they will learn a lot about how policy is made.

A similar program places scientists in prominent media outlets like the *Chicago Tribune,* the *Los Angeles*

Times, Scientific American, and National Public Radio. Plenty of prominent science journalists have come from this program. But even when program participants' visit to the world of journalism is temporary, they are—again—spreading and gaining useful knowledge.

Congressman Boehlert said in his *Science* interview that scientists and engineers should practice more advocacy. People who doubt him, he said, should look at the campaign platforms of people running for Congress and see who puts research spending or other technical issues at the top of their priority lists. "I'll bet you won't find one," he said. "And that's a failure of the scientific community."[17]

In some ways, John A. Knauss exemplifies how an accomplished scientist or engineer can make a big difference to society. Knauss is an eminent oceanographer who first made his name with his studies of ocean currents. But he is also a former chief of the National Oceanic and Atmospheric Administration (NOAA), founding dean of the Graduate School of Oceanography at the University of Rhode Island, and founder of the NOAA's Sea Grant program. He has strong views on how scientists can contribute to better policy, and the differences in approach that hinder their contributions. He outlined some of them in a speech prepared for a 1996 meeting of the Oceanography Society.

For one thing, Knauss said, "scientists are often unable to provide simple, unambiguous answers; our conclusions are seldom without caveats." But, he went on, the public does not necessarily require "the kinds of evidence scientists require of one another in our professional journals before being prepared to take action. . . . Approximate answers are adequate in many cases."[18]

He cited several examples, including the Montreal Protocol, an international agreement phasing out the production of chlorofluorocarbons, chemical compounds blamed for the depletion of Earth's protective ozone layer. The treaty came into force in 1989 even though, as Knauss said, "there was still some uncertainty as to the reason for the ozone hole."

While political action can take place amid some scientific uncertainty, uncertainty can also be used (or even drummed up) as an excuse to avoid action. That was the case with climate change, he said, and it was possible in part because "as scientists we have failed to communicate the severity of the problem." Scientific reports on the issue, he said, "are addressed to the mind and not to the heart."[19]

If you decide to speak out, you must be prepared for push-back. For example, Daniel Pauly, the fisheries expert at the University of British Columbia, has been highly and publicly critical of certain aquaculture and other industrialized fishing practices. He tells me there are entire Web sites devoted, as he puts it, "to showing what a complete idiot I am." Such personal attacks are distressing, even if you are accustomed to a certain amount of technical rough-and-tumble and even if you are confident you are correct.

There is no question that a life in public carries high costs. But scientists who speak out should not be punished by their own community. As Pauly put it in 2005, when he accepted the International Cosmos Prize, the notion that public engagement compromises scientific objectivity is "a tool" lobbyists and politicians have used to twist "the results of our work to fit their agenda."

The argument that engagement cannot coexist with technical objectivity "is never invoked in medicine," Pauly said. "Indeed, passionate engagement for the patients, against disease-causing agents is not only the norm, but also an essential element of doctors' professional ethics."[20]

15

Other Venues

In 2007, a graduate student in my Harvard seminar told me he was working for one of the candidates in the race for president. I was delighted—not because I thought his candidate was the best, but because it was so refreshing to see someone from academic research involved in politics. This kind of involvement, once common, is becoming more and more rare. And that is too bad. When scientists enter the political arena they can make a big difference.

That was the case in 2006 in Ohio, where scientists at Case Western Reserve University, Ohio State University, and elsewhere entered the political fray in hopes of thwarting the reelection of a member of the state board of education who favored including "intelligent design," a creationist theory, in Ohio public-school biology classes.

First, the group of scientists persuaded a candidate with sound science views to run against the other candidate. At Case, the effort's leaders produced an open letter, ultimately signed by almost all scientists on the university's faculty, urging voters "to stand up for quality science education" by supporting their candidate—and they made sure local news outlets heard about it.

Their candidate won handily.

Another organization, Scientists and Engineers for America, or SEA, aims to support candidates nationwide, whatever their party, who have what the group thinks is the right idea on science-related issues. And a group called Scientists and Engineers for Change entered presidential-race politics in 2004, sending scientists to give speeches arguing that the Bush administration had distorted science.

Still another group of scientists, Defend Science, was organized to oppose what it called "particular Christian fundamentalist 'moral codes' . . . increasingly imposing restrictions on what kinds of questions can be investigated by scientists and what kinds of answers scientists can come up with." Their litany of abuses includes thwarted research on stem cells, blocking or distorting findings on the prevention of HIV infection, and barring government scientists from even mentioning terms like *global warming.*[1] Defend Science has a division for working scientists and one for science students.

Meanwhile, students and postdocs at the University of Washington and the Fred Hutchinson Cancer Research Center in Seattle started the Forum on Science Ethics and Policy (FOSEP) to increase the dialogue between scientists and policymakers. It aims not just to bring science to the public but also to work "for the benefit of scientists, who may have a limited grasp of how the public sees their work."[2]

Not everyone thinks this kind of thing is a great idea.

In an editorial, the journal *Nature* warned that one of the dangers of winning the Nobel Prize is that people attempt to enlist you for all sorts of causes.[3] It particularly cited Scientists and Engineers for America and its opposition to Bush science policies, though "there is little doubt that US federal science has suffered under Bush," the editors wrote. By engaging in partisan behavior, the journal warned, scientists risk "seeming to be self-interested, grant-obsessed, and out of touch."

Actually, I think the reverse is true. It is remaining at the bench when times call for action that defines researchers as self-obsessed. As Burton Richter, a Stanford physicist, Nobel laureate, and founder and board member of SEA wrote in response to the *Nature* editorial, the organization's aim "is to make available to society at large the evidence-based science relating to critical issues facing us all." He added, "We hope both to draw attention to underappreciated science issues and provide the advocacy necessary to get things done—not along party-political lines but scientifically."[4]

David Baltimore, who became president of the American Association for the Advancement of Science in 2007, agrees. In his candidacy statement he said that, with science under increasing attack, "it's never been more important for scientists to take a role in public life."[5]

But even if you don't want to take so public a political role, there are other options to explore.

If you are a graduate student in science or engineering, you could consider participating in one of the fellowship programs run by the American Association for the Advancement of Science that send scientists to newsrooms and government offices.

Leon Lederman, the Nobel-laureate physicist and former director of the Fermi National Accelerator Laboratory, or Fermilab, has proposed a kind of "tax" on young researchers—a requirement that they spend some time helping high school teachers do a better job teaching science. Until such a tax is imposed, if it ever is, there is (in theory, anyway) nothing stopping you from volunteering to do something like this yourself.

If you are at an academic institution, you can urge colleagues to pay more scholarly attention to the ways in which science, technology, and policy overlap. As Sheila Jasanoff put it in online material describing a lecture series she launched at Harvard in 2006, "Within the academy, questions of science, technology and society often get short shrift because they belong at once to everybody and nobody." The sweeping societal transformations science and technology produce call for "new, multidisciplinary approaches," she wrote.

In his presidential address to the 2007 meeting of the American Association for the Advancement of Science, John Holdren urged his colleagues to "tithe 10 percent of your professional time and effort" to working to improve the human condition.

You can organize a Café Scientifique, or science café, where members of the public can gather, have a drink, and hear an interesting presentation on a scientific or technical topic. This phenomenon originated in the United Kingdom and is spreading around the world. "The format of the evening is simple," the UK Café Scientifique network says. First a short talk, usually "just words," no audiovisual or PowerPoint accoutrements.

Then questions and conversation between the presenter and the audience.

The science café format works because, first, it does not cost any money. People pay their own way. Also, cafés and restaurants like to host them because they usually draw enthusiastic crowds ready to eat and drink and enjoy an interesting talk. In fact, they have become so successful that the WGBH Educational Foundation in Boston and Sigma Xi, the scientific honorary society, have established a Web site *(www.sciencecafes.org)* to promote the movement in the United States.

Students at Harvard Medical School have organized something similar, a program called Science in the News. This organization holds regular discussions and presentations on topics in the news, usually because they are at the intersection of science and policy.

The Princeton Plasma Physics Laboratory takes another approach, offering free Saturday morning lectures, January through March, aimed at high school students but open to all. In 2008, topics ranged from the size of the universe to wound healing to the way rumors spread on Facebook.

In New York City in the spring of 2008, the World Science Festival offered dramas, lectures, story slams, and other events built around a scientific theme. The festival, organized by Brian Greene, a physicist at Columbia University, and his wife, Tracy Day, a former ABC television producer, offered a week of events, and most of them were sold out. Overall, the festival was such a hit its organizers hope to work with the city to run comparable events throughout the year.

Or you can voyage to SciLands, an archipelago of science islands in the virtual online world Second Life. Anthony Crider, an associate professor of physics at Elon University in North Carolina, has taken his students there. Crider, a cofounder of SciLands, devotes his site to science education and outreach. He is testing the environment as a teaching tool in his introductory astronomy courses.

In a talk at an AAAS forum in May 2008, Crider described how he discovered Second Life, concluded it was too preoccupied with gambling and sex, "and decided to try to do something constructive with it."[6]

In part because he did not have a planetarium at Elon, he built one online in Second Life. But pretty soon, he said, "something went up in the backyard, a female vampire porn shop. So I had to transplant it."

He approached Linden Labs, the proprietor of Second Life, and eventually he found himself tied in with the International Spaceflight Museum, the Exploratorium science museum in San Francisco, the University of Denver, the weekly National Public Radio show *Science*

Friday, NASA, and other organizations. Because "I did not want to have to deal with neighbors I did not like," Crider said, he wrote policy guidelines limiting who can build things on the site and so on.

Not all ideas work well in SciLands, Crider says. For example, he said NASA constructed an online classroom where people can watch PowerPoint presentations. "That does not work," he said. But elsewhere on the site you can visit the University of Denver's observatory—even its living quarters. Or you can stand on the surface of Mars and watch a lander descend. Or you can visit one of Crider's sites and walk on the surface of the Moon, "leaving little Moonprints as you walk around."

You could even make movies, like Randy Olson, who earned his PhD in biology at Harvard and won tenure at the University of New Hampshire before he decided he wanted to tell people about science on film, not in class. He made a number of small pieces before taking on the debates over creationism, in his 2006 documentary *Flock of Dodos* (the dodos are not necessarily who you think they are), and climate change, in the 2008 film *Sizzle: A Global Warming Comedy.* His movies are deliberately breezy and funny—some all-too-serious scientists might say they are a bit too funny. But as Al Gore showed with *An Inconvenient Truth,* film can teach people a lot.

Deciding to be a filmmaker is of course a life-changing decision. If you do not want to make films yourself, you could agree to be in someone else's movie, or on their television program. Or you could post something on YouTube. As Kate McAlpine, the Large Hadron Collider rapper, wrote after her YouTube success, "It would be nice if scientists who write about their work on blogs or craft articles for an outreach Web site could see those efforts valued on an equal footing with the hours they spend coding software, for instance." She added, "We need to keep putting information out at a level that people without specialized training can understand, appreciate, and maybe even dance to."[7]

Or you could sing and dance. Literally.

In 2008, the University of Rhode Island's honors colloquium concluded with *It's a Shore Thing,* a cabaret-style performance that included "Song of the Nile," about the relationship between politics, culture, and natural resources; "Be Careful," a song about invasive species; and "Ice Is Nice," a rap about the way the poles function as earthly air conditioners.

One of the creative forces behind this work at URI is Judith Swift, professor of communications studies, who began writing "coastal cabarets" at the university's Graduate School of Oceanography in the 1980s. She has explored topics as diverse as the Big Bang and undersea manganese nodules. Many of the songs are based on research conducted by members of the university faculty.

Even if the footlights are not in your future, these are efforts we all should applaud. When you encounter others embarking on these kinds of creative scientific ventures, don't denigrate their efforts. Cheer them on.

Here is what Neal Lane, the Rice University physicist and White House science adviser in the Clinton administration, has to say:

> I believe that the new definition of leadership for the science community must include a civic persona. Each of us will find our own path for this role . . . [I]n the past, science and technology have provided an important pathway to the American Dream—providing opportunities, fulfilling aspirations and promoting a better quality of life for all. Our task will be to take that message of the past and the promise science offers for the future to those we ask to pay a major part of the bill. Only then can we be assured that even in this new era of accountability, science is appreciated as a national investment and is able to continue to do its work.[8]

He wrote these words in an essay describing what happened when he was asked to give a talk at the Arlington, Virginia, Rotary Club—a talk he gave, he said, because he is "adamant about the importance of our message and getting it out to 'the people.'"

If you agree to give such a talk, consider it an important occasion. Think about it and prepare for it with the same thought and energy you would bring to a scientific presentation.

Should you use PowerPoint? I say no, because I think it tends to divide speakers from their audiences. If you must use it, limit your slides to compelling images that add something to your presentation rather than simply outlining what you are saying as they are being shown. In one of his early films, Olson, the biologist-turned-filmmaker, made this point by showing a typical scientist giving a typical presentation. The scientist spent 87 percent of his podium time looking at his own slides. Interact with your audience, not your slides.

Should you write out your whole talk in advance? Again, my advice is no. Outline your talking points and speak from them. Your talk will be more conversational and you will, again, forge a closer connection to your audience. When I give a talk, I print out my talking points in large type—fourteen-point type or even larger—so I will be able to read my notes easily from the podium. (I confess I am guilty of putting my glasses on and taking them off—but I am trying to break myself of this distracting habit.)

Remember that a lay audience will not have read the literature in your field and will be absorbing your information on the fly. Speak in simple declarative sentences and keep your subjects, verbs, and objects in order—and close to each other. Again, if you don't know what I am talking about here, invest in an English usage book.

Avoid anything in your talk that smacks of footnotes, but offer a brief description of the research that produced whatever findings you will be talking about.

Practice ahead of time. Record your talk and listen to or view the tape. Notice if you are speaking too fast or in a monotone. The more you practice, the more relaxed you will be on the day of the talk.

As with a press conference, if someone asks you a question you cannot answer, acknowledge your ignorance. If possible, suggest other sources for the information or, if practical, offer to provide it yourself. If someone in your audience becomes disagreeable, tell them you'll continue your conversation with him or her after the program ends. One way to cut disagreeable people short during the Q and A, especially when you are speaking in a large auditorium, is to have two microphones for questioners, one on each side of the hall. You can simply turn away from someone who is causing a problem and say "next question" to the person on the other side.

Many people find speaking in public daunting, but I can assure you from personal experience that it gets easier and easier with practice.

Conclusion

The chemist and novelist C. P. Snow, in a lecture in Cambridge, England, in 1959, spoke of "the two cultures," science and the humanities, and the gulf of mutual incomprehension he saw growing between them. "It is dangerous to have two cultures which can't or don't communicate," he said. "In a time when science is determining much of our destiny, that is, whether we live or die, it is dangerous in the most practical terms. . . . At present we are making do in our half-educated fashion, struggling to hear messages, obviously of great importance, as though listening to a foreign language."[1]

If anything, Snow's message is truer today than it

was then. In an editorial in the journal *Science,* Alan I. Leshner, then president of the American Association for the Advancement of Science, summed up the reasons, and why, as a result, clear communication of science and technology is more crucial now than ever.

Widely publicized instances of financial conflict of interest and fraud have tarnished the reputation of science as an institution, he wrote. Worse, there seems to be a growing array of issues in which the scientific position challenges—or is said to challenge—deeply held religious or political beliefs. Among these issues are evolution and climate change and stem cell research. "The ensuing tension threatens to compromise the ability of the scientific enterprise to serve its broad societal mission and may weaken societal support for science," Leshner wrote.[2]

Though Americans have tremendous respect for the ability of engineers and scientists to solve important problems and answer important questions, polls indicate many of us worry that technology moves too fast, and that its benefits blind us to important spiritual concerns.

Perhaps that thinking was what motivated Leon Kass, who chaired the President's Council on Bioethics from 2001 to 2005, in his opposition to the sexual revolution, abortion rights, and human embryonic stem cell research. Kass is among those who believe that advances in technology bring with them questions of spiritual and moral values that technology cannot cope with. This view is widespread, and for good reason.

As Vartan Gregorian, president of the Carnegie Corporation and former president of Brown University once put it, "Kass's concerns are especially relevant today, when we are awash in information, buffeted at every turn by data, breaking news, bulletins, faxes and email spam. The info-glut makes it much harder to integrate and give coherence to knowledge. Humanity has always craved meaning and wholeness, and when people do not have the ability or the knowledge to separate fact from fiction, to question deeply, to integrate knowledge, or to see coherence and meaning in life, they feel a deeply unsettling emptiness at the core of their lives."[3]

By explaining their work—and in particular, what motivates them to do their work—scientists and engineers can help the public understand and deal with what feels like a chaotic rush of technological change. In the absence of this understanding, though, Gregorian writes, members of the public may turn to religious "ideologies du jour," cults, or other "Cliffs Notes and catechisms" that offer their own kind of guidance.

Encouraging scientists and engineers to speak out in this kind of an environment will be difficult. For one thing, research as an institution needs to change, so that experts who take the time and make the effort to communicate with the public and participate in public discourse are rewarded for it. At a bare minimum, they should not be punished for it. As Leshner wrote, "that will entail putting public outreach efforts among the metrics used to decide promotion and tenure."[4]

If these and other exhortations reach their targets, there will be new demand for wider participation by scientists and engineers in the wider world. How to meet it?

Some, like Bruce Alberts, the editor of *Science,* call for new options in graduate education that would "prepare a student to become a professional policy analyst, a science education researcher, a science-oriented journalist, or a science curriculum specialist in a school district, for example."[5] In July 2008, the National Academy endorsed master's degree–level training for researchers interested in policy careers. But Alberts noted that these programs can suffer from both their separation from standard doctoral programs and their "limited capacity."

My own experience tells me there are many graduate students in technical fields who would be interested in pursuing such a course. But my experience also tells me they face opposition from their advisers and other senior researchers—and that this opposition will be hard to overcome.

As a society we need to adopt a broader view of what it means for researchers to fulfill their obligations to society. In my view, it is not enough for them to make findings and report them in the scholarly literature. As citizens in a democracy, they must engage, and not just when their funding is at stake.

For example, I wish researchers, in any scientific field, would find out what students in their towns are learning about evolution. And if they are not happy with what the schools are teaching, I wish they would speak out about it. When they see science being obviously misused in *any* way, I wish they would speak out. I wish in their ordinary lives they would help educate their friends and neighbors about how science works.

Leshner calls this a "glocal" approach: "taking a global issue and making it meaningful on a local level."[6] For example, he writes, scientists might recruit their non-scientist friends and neighbors to promote science funding to decision makers.

John Edward Porter, a Washington lawyer and former Republican congressman, gave his audience a similar message at a forum held in May 2008 by the American Association for the Advancement of Science. Porter, who as a congressman headed a committee that dealt with appropriations for the National Institutes of Health and the Centers for Disease Control and Prevention, told his audience to think routinely about how they can advance the cause of science by speaking out in public and to the government.

"I am talking about a lot more than voting on November 4th and paying your dues to AAAS," he said.[7] Speaking six months before the 2008 presidential election, he told his audience they should be organizing to put forward candidates for science adviser for the next president *and* for the forty or fifty major science-related positions in the executive branch. And they should be agitating for the removal of factors—like low salaries and needlessly restrictive personnel policies—that discourage scientists and engineers from participating in government service.

Porter said his listeners should be logging onto sites that have guides for voters, like Research!America, the organization he now heads, to see how their political officials have responded to questionnaires on science issues and other matters. If there are no responses, he said, they should call or write to ask why.

Even in political off years, he went on, researchers should offer their assistance to political figures, at all levels of government. "Ask yourself," he urged his AAAS audience, "wouldn't it be wonderful if all the candidates had science advisors or science advisory committees?" If you think the answer is yes, Porter said, "tell them you would like to be their science advisor or serve on his or her science advisory committee. If they say they don't have one, tell them you will create one for them. Press to put science into their message to the voters." Fewer than 3 percent of members of the U.S. House and Senate have advanced technical training, he went on. "They need all the help they can get."

Porter, who described himself as one of a dying breed of moderate Republicans, urged his audience not to focus on one political party. In the first place, he said, "you want both parties committed to science." And, he added, in politics things change quickly, so a politician who seems invincible one day may be out of the picture the next. His advice: spread your bets.

Write op-eds on science funding and other issues, he urged. "Take your newspaper's science reporter out to lunch. Make a speech at your town's service club. Reach out to the public. Invite your Representatives and Senators to campus to see your research. I guarantee they will be fascinated. And even if they don't come their staff will come, and they are just as important."

By this point, Porter was on a roll. The moderator of

his panel at the AAAS forum informed him that he had run over his time, but he would not be stopped. "This message has to be heard!" he said, and he kept talking.

"Scientists are by every measure the most respected people in America," he said. "They are listened to. But if the public and policymakers never hear your voices, never see . . . science, never understand its methods, the chance of its being high on the list of national priorities will be very low."

Porter concluded, "You can sit on your fingers or you can go outside your comfort zone and get into the game and make a difference for science. . . . [N]either we nor AAAS, nor any other group can do it all for you. Science needs you. Your country needs you. America needs you . . . fighting for science!"

The hall erupted in wild applause. I could only hope the people clapping would act on the advice.

This book has focused on the practical, political, and policy reasons why it is important for scientists and engineers to engage more vigorously in the public life of the nation. But there is another, far more important reason.

Brian Greene, the Columbia University physicist and author of *The Elegant Universe,* cited it on the op-ed page of the *New York Times* when he urged his readers to, as the headline put it, "Put a Little Science in Your Life," not because it is useful or good for you or important to the progress of society. Do it, he wrote, because the research enterprise offers us optimism in a confusing world full of disappointment and bad actions.

"Science is a process that takes us from confusion to understanding in a manner that's precise, predictive and reliable—a transformation, for those lucky enough to experience it, that is transformative and emotional," he wrote.[8] I am reminded of this whenever I think of the 200-inch telescope on Palomar Mountain in California. For scientists and engineers, the telescope is notable because for years it was the largest optical instrument in the world, capable of amazing observations. For me, it is notable for what it tells us about the relationship between people and research.

Though the telescope did not begin operation until 1948, its mirror, cast at the Corning Glass Works in Corning, New York, traveled to California by train in 1936. The trip took sixteen days. All across the country, people stood along the tracks to watch it go by.

I think I know why. Things were not going so well in 1936. The country was still in the grip of the Depression and totalitarianism was on the march in Europe and Asia. But the engineers who made that mirror and the astronomers who would use it embodied the aspirations of our research, and our species' insatiable curiosity about the natural environment in which we live. I think the people who watched along the tracks for the mirror to pass by wanted to partake in the great adventure of research. Then, as now, that kind of work offers all of us a beacon of hope.

Notes

Acknowledgments

Suggested Reading

Bibliography

Index

Notes

1. An Invitation to Researchers

1. Personal communication.

2. Richard Gallagher, "Wanted: Scientific Heroes," *Scientist,* July 18, 2005, 6.

3. Bridging the Sciences Survey, 2006, Charlton Research Company for Research!America, described by Mary Woolley at the American Association for the Advancement of Science (AAAS) Forum on Science and Technology Policy, Washington, DC, May 8, 2008.

4. Donald Kennedy, remarks at the American Academy of Arts and Sciences, Cambridge, MA, February 13, 2008.

2. Know Your Audience

1. Presentation by Josh Tenenbaum at Medical Evidence Boot Camp, Knight Science Journalism Fellowships at MIT, Cambridge, MA, 2003.

2. Bridging the Sciences Survey, 2006, Charlton Research Company for Research!America, described by Mary Woolley at the AAAS Forum on Science and Technology Policy, Washington, DC, May 8, 2008.

3. Robert Park, *Voodoo Science: The Road from Foolishness to Fraud* (New York: Oxford University Press, 2000), 35–36.

4. National Science Foundation, *Science and Engineering Indicators 2008* (Arlington, VA: National Science Board, 2008).

3. The Landscape of Journalism

1. Scott Gant, *We're All Journalists Now: The Transformation of the Press and Reshaping of the Law in the Internet Age* (New York: Free Press, 2007).

2. Brian Lambert, "Twin Cities Editor Planning Online Daily," *New York Times,* August 27, 2007, C5.

3. Thomas Patterson, *Young People and News,* a report from the Joan Shorenstein Center for the Press, Politics and Public Policy, John F. Kennedy School of Government, Harvard University, July 2007.

4. Ibid., 5.

5. Stanley Walker, *City Editor* (1934; repr., Baltimore: Johns Hopkins University Press, 1999), 44.

4. Covering Science

1. Cristine Russell, "Covering Controversial Science: Improving Reporting on Science and Public Policy," Working Pa-

per 2006-04, Joan Shorenstein Center for the Press, Politics and Public Policy, Harvard University, 2006.

2. Cornelia Dean, "Rousing Science Out of the Lab and into the Limelight," *New York Times,* November 11, 2003, F10.

5. The Problem of Objectivity

1. Daniel Yankelovich, "Winning Greater Influence for Science," *Issues in Science and Technology,* Summer 2003. Cornelia Dean, "Rousing Science Out of the Lab and into the Limelight," *New York Times,* November 11, 2003, F10.

2. Personal communication.

3. Lawrence Krauss, remarks in panel discussion organized by *Scientific American,* Columbia University, October 26, 2006.

4. Personal communication.

5. Naomi Oreskes, "Beyond the Ivory Tower: The Scientific Consensus on Climate Change," *Science* 306 (December 3, 2004): 1686.

6. The Scientist as Source

1. Roald Hoffmann, "The Metaphor, Unchained," *American Scientist* 94 (September-October 2006).

2. Scott Morgan and Barrett Whitener, *Speaking about Science: A Manual for Creating Clear Presentations* (Cambridge: Cambridge University Press, 2006).

3. "Talk to the Newsroom: Graphics Director Steve Duenes," *www.nytimes.com,* posted February 25, 2008.

4. Personal communication.

5. Centers for Disease Control and Prevention, "Guide-

lines for Investigating Clusters of Health Events," *Morbidity and Mortality Weekly Report* 39 (July 27, 1990): 5.

6. "Why Giuliani's Campaign Was a Flop," *The Week,* February 15, 2008, 14.

7. *Branzburg v. Hayes,* 408 U.S. 665 (1972).

8. Richard Hayes and Daniel Grossman, *A Scientist's Guide to Talking with the Media: Practical Advice from the Union of Concerned Scientists* (New Brunswick, NJ: Rutgers University Press, 2006), 25.

9. Quoted in *Eos* 68, no. 28 (July 10, 2007): 289.

7. Public Relations

1. Earle M. Holland, "Working with Information Specialists," in Melissa K. Welch-Ross and Lauren G. Fasic, eds., *Handbook on Communicating and Disseminating Behavioral Science* (Thousand Oaks, CA: Sage Publications, 2007), 203.

2. Personal communication.

3. American Academy of Arts and Sciences panel on science communication, Cambridge, MA, February 14, 2008.

4. Personal communication

5. Holland, "Working with Information Specialists," 206.

6. Ibid., 213.

7. Ibid., 214.

8. Telling Stories on Radio and TV

1. Jennifer Jacobson, "Loving the Limelight," *Chronicle of Higher Education,* April 21, 2006, 13.

2. Frank Kauffman, remarks at the Aldo Leopold Leadership Program, West Cornwall, CT, June 2008.

3. Personal communication.

9. Telling Science Stories Online

1. "Large Hadron Rap," posted by Alpinekat, has had millions of hits on YouTube.

2. "This Ain't No Jive, Particle Physics Rap Is a Hit," Associated Press, September 1, 2008.

3. Kate McAlpine, "Commentary: Rapping Physics," *Symmetry* 5, no. 5 (November 2008).

4. Personal communication.

5. Steve Outing and Laura Ruel, "The Best of Eyetrack III," *www.poynterextra.org/eyetrack2004.*

6. You can register a Web site at *www.networksolutions.com.*

7. See *www.shiftingbaselines.org* and *www.climatecentral.org.*

8. Sheril Kirshenbaum, remarks at the AAAS Forum on Science and Technology Policy, Washington, DC, May 2008.

9. Scott Gant, *We Are All Journalists Now: The Transformation of the Press and Reshaping of the Law in the Internet Age* (New York: Free Press, 2007). Sarah Boxer, "Blogs," *New York Review of Books,* February 14, 2008.

10. Alison Ashlin and Richard J. Ladle, "Environmental Science Adrift in the Blogosphere," *Science* 312 (April 14, 2006): 201.

11. E. Jean Carroll, "Ask E. Jean," *Elle,* August 28, 2007, 384.

12. Paul Boutin, "So You Want to Be a Blogging Star?" *New York Times,* March 20, 2008, C8.

13. Clark Hoyt, "Civil Discourse, Meet the Internet," *New York Times,* November 4, 2007, WK14.

14. Sharon Otterman, "Haste, Scorned: Blogging at a Snail's Pace," *New York Times,* November 23, 2008, 10.

15. Ibid.

10. Writing about Science and Technology

1. Donald Kennedy, "A New Year and Anniversary," *Science* 307 (January 7, 2005): 17.

2. David Damrosch, "Trading Up with Gilgamesh," *Chronicle of Higher Education,* March 9, 2007, B5.

3. Vernon Booth, *Communicating Science: Writing a Scientific Paper and Speaking at Scientific Meetings* (Cambridge: Cambridge University Press, 2000).

4. Roald Hoffmann, "The Metaphor, Unchained," *American Scientist* 94 (September-October 2006), 407.

5. See, e.g., Philip Corbett, "Smoothing the Rough Spots," After Deadline, December 16, 2008, on the Times Topics blog, *topics.blogs.nytimes.com/2008/12/16/smoothing-the-rough-spots.*

6. Personal communication.

7. *Science Writer,* Spring 1992, 8.

8. Peter Elbow, *Writing with Power: Techniques for Mastering the Writing Process* (New York: Oxford University Press, 1998).

9. Hoffmann, "The Metaphor, Unchained," 407.

10. Personal communication.

11. The Editorial and Op-Ed Pages

1. Personal communication.

2. Felicia Ackerman, "Dealing with Dementia," Letters, *New York Times,* November 9, 2004.

3. Felicia Ackerman, "Kinsey, the Moralizer," Letters, *New York Times,* October 10, 2004.

4. David Shipley, "And Now a Word from Op-Ed," *New*

York Times, February 1, 2004; *www.nytimes.com/2004/02/01/opinion.*

12. Writing Books

1. Susan Rabiner and Alfred Fortunato, *Thinking Like Your Editor: How to Write Serious Nonfiction—and Get It Published* (New York: W. W. Norton, 2002).

2. Rachel Toor, "The Care and Feeding of the Reader," *Chronicle of Higher Education,* September 14, 2007, C2.

3. James D. Watson, *Avoid Boring People: Lessons from a Life in Science* (New York: Oxford University Press, 2007), 236–237

4. Rabiner and Fortunato, *Thinking Like Your Editor.*

5. Personal communication.

6. Personal communication.

7. Georgina Ferry, "A Scientist's Life for Me," *Nature* 16 (October 2008): 871.

8. Ibid.

9. Ibid.

13. On the Witness Stand

1. *General Electric Co. v. Joiner,* 522 U.S. 136 (1997).

2. National Science Foundation, *Biology and Law: Challenges of Adjudicating Competing Claims in a Democracy* (Arlington, VA: National Science Foundation, 1995), 10ff.

3. Ibid., 11.

4. *Frye v. U.S.*, 54 App. D.C. 46, 293 F 1013 No. 3968.

5. Cynthia Crossen, *Tainted Truth: The Manipulation of Fact in America* (New York: Touchstone, 1996), 196.

6. Gina Kolata, "Panel Confirms No Major Illness Tied to Implants," *New York Times*, June 21, 1999, A1.

7. *Daubert v. Merrell Dow Pharmaceuticals*, 509 U.S. 579 (1993).

8. *General Electric Co. v. Joiner*.

9. Personal communication.

10. Robert Livingston, speaking at a meeting of Pew Marine Fellows, Blaine, WA, October 16–19, 2003.

11. Crossen, *Tainted Truth*, 199.

12. The First Amendment Project, which is a nonprofit advocacy organization, offers information on its Web site, *www.thefirstamendment.com*.

13. Peter Shelley, speaking at the meeting of Pew Marine Fellows, Blaine, WA, October 16–19, 2003.

14. Livingston, at meeting of Pew Marine Fellows.

14. Making Policy

1. Sherwood Boehlert, remarks at the AAAS Forum on Science and Technology Policy, Washington, DC, May 3, 2007.

2. Ibid.

3. Daniel Sarewitz, "Liberating Science from Politics," *American Scientist* 94 (March-April 2006): 104.

4. Daniel Greenberg, *Science, Money, and Politics: Political Triumph and Ethical Erosion* (Chicago: University of Chicago Press, 2001), 5.

5. Boehlert, AAAS Forum on Science and Technology Policy.

6. Martin Rees, "Science: The Coming Century," *New York Review of Books*, November 20, 2008, 42.

7. Ibid.

8. Editorial, "Spread the Word," *Nature* 451 (January 10, 2008).

9. Kate von Holle, "AGU Position Statements: Evolution and Nuclear-Test-Ban Treaty," *Eos* 89, no. 3 (January 15, 2008): 24.

10. Timothy H. Dixon and Roy K. Dokka, "Earth Scientists and Public Policy: Have We Failed New Orleans?" *Eos* 89, no. 10 (March 4, 2008): 96.

11. Ibid.

12. "Sherwood Boehlert Interview: Explaining Science to Power: Make It Simple, Make It Pay," *Science* 24 (November 2006): 1228.

13. Ibid.

14. Linda Michaluk, remarks at the AAAS Forum on Science and Technology Policy, Washington, DC, May 2008.

15. Personal communication.

16. William G. Wells, Jr., *Working with Congress: A Practical Guide for Scientists and Engineers*, 2nd ed. (Washington, DC: AAAS, 1996)

17. "Sherwood Boehlert Interview: Explaining Science to Power."

18. John A. Knauss, "The Politics of Global Warming," remarks prepared for meeting of the Oceanography Society, Seattle, April 1, 1996.

19. Ibid.

20. Daniel Pauly, "An Ethic for Marine Science: Thoughts on Receiving the International Cosmos Prize," remarks at the 13th International Cosmos Prize ceremony, Osaka, October 18, 2005.

15. Other Venues

1. See "The Statement" on the Defend Science Web site, *www.defendscience.org.*

2. Jonathan Knight, "Students Set Up Forum to Debate Hot Topics," *Nature* 431 (September 23, 2004): 390.

3. Editorial, "Nobels in Dubious Causes," *Nature* 447 (May 24, 2007): 354.

4. Correspondence, "Nobel Laureates Know What They're Talking About," *Nature* (July 26, 2007).

5. Becky Ham, "Baltimore: Time for Scientists to Take Role in Public Life," *Science* 315 (January 26, 2007).

6. Anthony Crider, remarks at the AAAS Forum on Science and Technology Policy, Washington, DC, May 2008.

7. Kate McAlpine, "Commentary: Rapping Physics," *Symmetry* 5, no. 5 (November 2008).

8. Neal Lane, "The Arlington Rotary Club," *American Scientist* 84 (1996): 208.

Conclusion

1. C. P. Snow, *The Two Cultures* (Cambridge: Cambridge University Press, 1998), 98.

2. Alan I. Leshner, "Outreach Training Needed," *Science* 315 (January 12, 2007): 161.

3. Vartan Gregorian, "Grounding Technology in Both Science and Significance," *Chronicle of Higher Education,* December 9, 2005, B3.

4. Leshner, "Outreach Training Needed."

5. Bruce Alberts, "New Career Paths for Scientists," *Science* 320 (April 18, 2008): 289.

6. Alan I. Leshner, "'Glocal' Science Advocacy," *Science* 319 (February 15, 2008), 877.

7. John Edward Porter, remarks at the AAAS Forum on Science and Technology Policy, Washington, DC, May 2008.

8. Brian Greene, "Put a Little Science in Your Life," *New York Times*, June 1, 2008, WK14.

Acknowledgments

I am grateful to the scores of people who over the years have shared with me their expertise and instincts on the communication of science and technology to the lay public. Among them are:

Fellow journalists, including Paula Apsell, Bill Blakemore, Jeff Burnside, Gareth Cook, Steve Curwood, Chris Joyce, David Malakof, Joe Palca, Ken Weiss, and of course my past and present science writing colleagues at the *New York Times,* including Natalie Angier, Sandra Blakeslee, Warren E. Leary, Andrew C. Revkin, and Sheryl Gay Stolberg, and George Freeman and his colleagues in the legal department of the *Times*

Professional communicators, including Nancy Baron, Rick Borchelt, Earle Holland, Frank Kauffman, and Dennis Meredith

Public servants and scholars of science and others, including Alan Alda, Sherwood Boehlert, Vernon Ehlers, David Goldston, John Holdren, Sheila Jasanoff, John Knauss, Jon Miller, and William A. Wulf

I have been the beneficiary of their wisdom, but the errors in this book are mine.

I am also very grateful to the scientists, engineers, and graduate students who have over the years participated in my seminars, particularly the ones who give me feedback and keep in touch with me about their efforts to reach out to the public.

I especially want to thank the people who made those seminars possible, including Alan Lightman of the Massachusetts Institute of Technology, Andries van Dam of Brown University, David Farmer of the Graduate School of Oceanography at the University of Rhode Island, and Jackleen de La Harpe, the founding director of the Metcalf Institute for Environment and Marine Reporting, and also James Clem of the Harvard University Center for the Environment and James McCarthy, Alexander Agassiz Professor of Biological Oceanography at Harvard, who gave me my first chance to teach at the university.

I am grateful to my agent, James Levine, and to Michael Fisher, the editor at Harvard University Press, for their enthusiasm for the project. Jill Breitbarth, the jacket designer, and Anne Zarrella, Susan Abel, and Kate Brick at Harvard University Press provided valuable support. I am indebted to the proofreader, Michael Baker, too. Julie Hagen is a model of the diligent and astute editor; I was fortunate to have her help with my manuscript.

Finally I want to acknowledge the enormous debts I owe to Joseph Lelyveld, who as executive editor of the *New York Times* appointed me the newspaper's science editor, and to Daniel Schrag, who as director of the Harvard University Center for the Environment has made it possible for me to share my thoughts on communicating research with graduate students there.

Suggested Reading

Classics

These books should be on every researcher's reading list. My advice: add them to your collection and dip into them from time to time.

Kuhn, Thomas S. *The Structure of Scientific Revolutions.* Chicago: University of Chicago Press, 1996.

Merton, Robert K. *On Social Structure and Science.* Chicago: University of Chicago Press, 1996. See especially Chapter 20 ("The Ethos of Science"), Chapter 21 ("Science and the Social Order"), and Chapter 22 ("The Reward System of Science"). Remember that Merton was not writing about research as he thought it ought to be, but rather research as it was. Then contemplate how things have changed.

Snow, C. P. *The Two Cultures.* Cambridge: Cambridge University Press, 1998. This is Snow's 1959 Rede Lecture on the divide between technical elites and literary intellectuals. What he says is perhaps even more relevant today than it was then. The edition I have includes both his speech as delivered and the text as revised for publication later. There are not too many differences between them, but the differences are illuminating.

Reporting and Writing

Here are some useful books on reporting and writing. Some of them, like the stylebooks, deal with such nuts-and-bolts questions as which words to abbreviate and which to spell out, when to hyphenate, and so on. Two of them, *On Writing Well* and *Elements of Style,* are classics. All of them are useful.

Blum, Deborah, and Mary Knudson. *A Field Guide for Science Writers.* Oxford: Oxford University Press, 1997.

Booth, Vernon. *Communicating Science: Writing a Scientific Paper and Speaking at Scientific Meetings.* Cambridge: Cambridge University Press, 2000.

Cappon, Rene J. *The Associated Press Guide to Good Writing.* Stamford, CT: Thomson Learning/ARCO, 1982.

Clark, Roy Peter. *Writing Tools.* Boston: Little, Brown, 2006.

Elbow, Peter. *Writing with Power: Techniques for Mastering the Writing Process.* New York: Oxford University Press, 1998.

Friedman, Sharon M., Sharon Dunwoody, and Carol Rogers, eds. *Scientists and Journalists: Reporting Science as News.* New York: Free Press, 1986.

Goldstein, Norm, ed. *The Associated Press Stylebook.* New York: Basic Books, 2004.

Kalbfeld, Brad. *Associated Press Broadcast News Handbook.* New York: McGraw-Hill, 2001.

Kern, Jonathan. *Sound Reporting: The NPR Guide to Audio Journalism and Production.* Chicago: University of Chicago Press, 2008.

Kramer, Mark, and Wendy Call. *Telling True Stories: A Nonfiction Writer's Guide from the Neiman Foundation at Harvard University.* New York: Plume, 2007.

Rabiner, Susan, and Alfred Fortunato. *Thinking Like Your Editor: How to Write Serious Nonfiction—and Get It Published.* New York: W. W. Norton, 2002.

Rosenberg, Barry J. *Spring into Technical Writing for Scientists and Engineers.* Reading, MA: Addison-Wesley, 2005.

Siegal, Allan M., and William G. Connolly. *The New York Times Manual of Style and Usage.* New York: Three Rivers Press, 2002.

Strunk, William, and E. B. White. *Elements of Style,* 50th anniversary ed. Upper Saddle River, NJ: Longman, 2008.

Zinsser, William. *On Writing Well,* 30th anniversary ed. New York: Collins, 2006.

Reading Other Writers

There are many collections of writing on science and technology. Some of them come out every year. Here are four recent examples:

Groopman, Jerome, ed. *Best American Science and Nature Writing 2008.* Boston: Houghton Mifflin, 2008.

Nasar, Sylvia, ed. *Best American Science Writing 2008.* New York: Harper Perennial, 2008.

Thompson, Clive, ed. *The Best of Technology Writing 2008.* Digital Culture Books, 2008.

The Oxford Book of Modern Science Writing. Ed. Richard Dawkins. Oxford: Oxford University Press, 2008. This is a more sweeping collection, both in terms of the prominence of its collected authors and the dates at which they wrote.

If you want to take a more unorthodox approach, read some plays with scientific themes. They are useful guidance for would-be science writers, I think, because playwrights dealing with science themes must explain technical topics so clearly they can be understood by an audience listening to the explanations on the fly—and their writing must be dramatically engaging. Here are some examples—good stories, well told and accurate.

Auburn, David. *Proof.* London: Faber and Faber, 2001.

Edson, Margaret. *Wit.* London: Faber and Faber, 1999.

Frayn, Michael. *Copenhagen.* Norwell, MA: Anchor, 2000.

Parnell, Peter. *QED.* New York: Applause Books, 2002.

Stoppard, Tom. *Arcadia.* London: Faber and Faber, 1994.

The Visual Display of Information

The work of Felice Frankel and Edward Tufte is interesting not just because of their astonishing technical skills but also for the thoughtfulness, imagination, and wisdom they bring to the visual display of scientific or technical information. Dipping into their books won't make you their equal in presenting your data—probably nobody can equal their work—but you will glean useful ideas about how to approach the task in your own lab.

Frankel, Felice. *Envisioning Science: The Design and Craft of the Science Image.* Cambridge, MA: MIT Press, 2002.

Frankel, Felice, and George M. Whitesides. *On the Surface of Things: Images of the Extraordinary in Science.* Cambridge, MA: Harvard University Press, 2007.

Tufte, Edward. *Envisioning Information.* Cheshire, CT: Graphics Press, 1990.

———. *The Visual Display of Quantitative Information.* Cheshire, CT: Graphics Press, 2001.

Dealing with the Public and Public Officials

Hayes, Richard, and Daniel Grossman. *A Scientist's Guide to Talking with the Media.* New Brunswick, NJ: Rutgers University Press, 2006.

Pielke, Roger A., Jr. *The Honest Broker: Making Sense of Science in Policy and Politics.* Cambridge: Cambridge University Press, 2007.

Welch-Ross, Melissa K., and Lauren G. Fasig, eds. *Handbook on Communicating and Disseminating Behavioral Science.* Los Angeles: Sage Publications, 2007.

Wells, William G., Jr. *Working with Congress: A Practical Guide for Scientists and Engineers,* 2nd ed. Washington, DC: American Association for the Advancement of Science, 1996.

Bibliography

Adam, Pegie Stark, et al. *Eyetracking the News: A Study of Print and Online Reading.* St. Petersburg, FL: The Poynter Institute, 2007.

Agin, Dan. *Junk Science.* New York: Thomas Dunne Books/St. Martin's Press, 2007.

Angell, Marcia. *Science on Trial: The Clash of Medical Evidence and the Law in the Breast Implant Case.* New York: W. W. Norton, 1996.

Bethell, Tom. *The Politically Incorrect Guide to Science.* Washington, DC: Regnery, 2005.

Booth, Vernon. *Communicating Science: Writing a Scientific Paper and Speaking at Scientific Meetings.* Cambridge: Cambridge University Press, 2000.

Crossen, Cynthia. *Tainted Truth: The Manipulation of Fact in America.* New York: Touchstone, 1996.

Faigman, David L. *Laboratory of Justice.* New York: Times Books, 2004.

———. *Legal Alchemy: The Use and Misuse of Science in the Law.* New York: W. H. Freeman, 1994.

Federal Justice Center. *Reference Manual on Scientific Evidence,* 2nd ed. Eagan, MN: West Group, 2000.

Federal Rules of Evidence. Louisville, CO: National Institute for Trial Advocacy, 2003.

Foster, Kenneth R., and Peter W. Huber. *Judging Science: Scientific Knowledge and the Federal Courts.* Cambridge, MA: MIT Press, 1997.

Frank, Marcie. *How to Be an Intellectual in the Age of TV.* Durham, NC: Duke University Press, 2005.

Friedman, Sharon M., et al., eds. *Communicating Uncertainty.* Philadelphia: Lawrence Erlbaum Associates, 1999.

Gant, Scott. *We're All Journalists Now: The Transformation of the Press and Reshaping of the Law in the Internet Age.* New York: Free Press, 2007.

Gigerenzer, Gerd. *Calculated Risks: How to Know When the Numbers Deceive You.* New York: Simon and Schuster, 2002.

Golan, Tal. *Laws of Men and Laws of Nature.* Cambridge, MA: Harvard University Press, 2004.

Goldstein, Tom. *Journalism and Truth: Strange Bedfellows.* Evanston, IL: Northwestern University Press, 2007.

Greenberg, Daniel S. *Science, Politics, and Money: Political Triumph and Ethical Erosion.* Chicago: University of Chicago Press, 2001.

Hartz, Jim, and Rick Chappell. *Worlds Apart: How the Distance between Science and Journalism Threatens America's Future.* Nashville: First Amendment Center, 1997.

Hayes, Richard, and Daniel Grossman. *A Scientist's Guide to Talking with the Media: Practical Advice from the Union of Concerned Scientists.* New Brunswick, NJ: Rutgers University Press, 2006.

Jacoby, Susan. *The Age of American Unreason.* New York: Pantheon Books, 2008.

Jasanoff, Sheila. *The Fifth Branch.* Cambridge, MA: Harvard University Press, 1990.

———. *Science at the Bar: Law, Science, and Technology in America.* Cambridge, MA: Harvard University Press, 1995.

Krimsky, Sheldon. *Science in the Private Interest: Has the Lure of Profits Corrupted Biomedical Research?* Lanham, MD: Rowman and Littlefield, 2004.

Kuhn, Thomas S. *The Structure of Scientific Revolutions.* Chicago: University of Chicago Press, 1996.

Marshall, Stephanie Pace, et al., eds. *Science Literacy for the Twenty-first Century.* Amherst, NY: Prometheus Books, 2003.

Merton, Robert K. *On Social Structure and Science.* Chicago: University of Chicago Press, 1996.

———. *The Sociology of Science.* Chicago: University of Chicago Press, 1973.

Morgan, Scott, and Barrett Whitener. *Speaking about Science: A Manual for Creating Clear Presentations.* Cambridge: Cambridge University Press, 2006.

National Science Foundation. *Science and Engineering Indicators 2006.* Arlington, VA: National Science Board, 2006.

Park, Robert. *Superstition: Belief in the Age of Science.* Princeton, NJ: Princeton University Press, 2008.

———. *Voodoo Science: The Road from Foolishness to Fraud.* New York: Oxford University Press, 2000.

Pielke, Roger A., Jr. *The Honest Broker: Making Sense of Science in Policy and Politics.* Cambridge: Cambridge University Press, 2007.

Rabiner, Susan, and Alfred Fortunato. *Thinking Like Your Editor: How to Write Serious Nonfiction—and Get It Published.* New York: W. W. Norton, 2002.

Ropeik, David, and George Gray. *Risk.* Boston: Houghton Mifflin, 2002.

Slovic, Paul. *The Perception of Risk.* London: Earthscan, 2002.

Snow, C. P. *The Two Cultures.* Cambridge: Cambridge University Press, 1998.

Watson, James D. *Avoid Boring People: Lessons from a Life in Science.* New York: Oxford University Press, 2007.

Wells, William G., Jr. *Working with Congress: A Practical Guide for Scientists and Engineers,* 2nd ed. Washington, DC: AAAS, 1996.

Index